LabVIEW
Digital Signal
Processing

LabVIEW Digital Signal Processing

and Digital Communications

Cory L. Clark
Motorola

McGraw-Hill

New York   Chicago   San Francisco   Lisbon   London   Madrid
Mexico City   Milan   New Delhi   San Juan   Seoul
Singapore   Sydney   Toronto

The McGraw·Hill Companies

Cataloging-in-Publication Data is on file with the Library of Congress

Copyright © 2005 by The McGraw-Hill Companies, Inc. All rights reserved. Printed in the United States of America. Except as permitted under the United States Copyright Act of 1976, no part of this publication may be reproduced or distributed in any form or by any means, or stored in a data base or retrieval system, without the prior written permission of the publisher.

1 2 3 4 5 6 7 8 9 0   DOC/DOC   0 1 0 9 8 7 6 5

ISBN 0-07-144492-0

The sponsoring editor for this book was Stephen S. Chapman and the production supervisor was Sherri Souffrance. It was set in Century Schoolbook by International Typesetting and Composition. The art director for the cover was Margaret Webster-Shapiro.

Printed and bound by RR Donnelley.

McGraw-Hill books are available at special quantity discounts to use as premiums and sales promotions, or for use in corporate training programs. For more information, please write to the Director of Special Sales, McGraw-Hill Professional, Two Penn Plaza, New York, NY 10121-2298. Or contact your local bookstore.

To Z, who always shows me what's important; to my parents, who started me out right; and to my sister Holly, who taught me to read

Contents

Preface

About This Book

This is not a book about how to use LabVIEW or even a book on learning digital signal processing (DSP). Instead it is more of a practical guide on how to enable LabVIEW to tackle some real-world DSP and communication problems. This book assumes that the reader has a good grasp of many of the complex issues encountered in DSP and digital communications and also is at least skilled enough in LabVIEW to build a VI. When necessary, the book will dive into the heart of signal processing topics and their implications will be explored. Certain topics will be explained in enough detail so that the reader will know there is no hand waving or mystery involved. This material is meant to bridge the gap between obtaining theoretical knowledge and actually exercising that knowledge. LabVIEW provides us with an excellent set of tools for examining all sorts of DSP and digital communication topics. Its graphical nature allows us to quickly and efficiently get to the core of a communication problem without all the overhead that generally accompanies a digital communication system. This book will start out at the beginning of the DSP realm—sampling a signal. The intermediate chapters will cover some basic building blocks and the final chapters will put it all together as a digital communication system.

A lot of signal processing books start out describing what a discrete time sequence is, the advantages of DSP over analog methods, and the like. This book skips all that and assumes that you already know enough about DSP to get started and you probably have some very good references regarding where to go when you do not understand something. Instead this book focuses on putting that DSP knowledge to work using LabVIEW. Also, at the end of each chapter is a list of references for the specific topics covered in that chapter. Of course the reader is encouraged to look at those references for any concept that is not quite clear. If your DSP is a little rusty, or if you are new to the topic, a good starting place would be to read *Understanding Digital Signal Processing* by Rick Lyons before moving to the more advanced texts such as *Discrete-Time Signal Processing* by Oppenheim and Schafer. The book by Lyons should give you a good intuitive feel for many complicated DSP subjects while the Oppenheim and Schafer book will give you all the gory details on how and why.

As with any subject, you can read about DSP all day long and not quite understand it until you actually put it into practice. Hopefully, after working your way through this book, you will not only get better at using LabVIEW, but your signal processing skills will be more instinctive. Most engineers and students are familiar with Matlab because it is the most common DSP simulation environment. But this book attempts to show that almost everything that Matlab can do, LabVIEW can do just as easily (and perhaps more easily). LabVIEW has two distinct areas where it excels over Matlab: (1) its graphical nature—you can look at what is going on, not just interpret words on a page—and (2) its interface to external hardware and instruments. LabVIEW combines these characteristics with some very useful built-in functions to perform all sorts of signal processing. All of the examples in this book are compatible with LabVIEW 7.0 express evaluation version. This software may be downloaded free of charge from the National Instruments website and the software will run for 30 days. All references and built-in VIs are included in the 7.0 evaluation version and are not guaranteed to work on any other version of LabVIEW. The only exceptions to this will be the special toolsets that National Instruments ships with some of their RF measurement hardware. These VIs will not be necessary to experiment with any of the VIs in this book, but certain functions may be mentioned for completeness.

Organization of the Book

This book tries to take the following approach—begin at the beginning and build strong fundamentals with each chapter building on the previous ones. Following that theme, the book is divided into three parts: Getting Started, Building Blocks, and finally, Building a Communication System. In this case, the beginning will be, "Why LabVIEW?" Here the book moves slowly through Chapters 1 and 2 covering the intricate details of actually acquiring a signal. Chapter 3 explores the LabVIEW spectral processing tools such as DFT and also touches on some of the impairments associated with DFT computation. Chapter 4 shows how to design digital filters in LabVIEW and Chapter 5 uses those filter design concepts in the context of multirate sampling. Next, some very useful signals are generated in Chapter 6, and we look at mapping of bits to symbols for building a modulated waveform. In Chapters 7 and 8 we build and evaluate our digital communication system. Finally, Chapter 9 reveals a few techniques for optimizing the speed of signal processing computations in LabVIEW.

Acknowledgments

You may not know this, but writing a book is hard work. You scrutinize every word and worry that you have said something completely wrong. Because of the monumental nature of researching, writing, organizing, editing, and publishing, no book is a single-handed effort. As such, I would like to acknowledge the following people for their part in helping me through this: Mark Goldberg and Stephen Shiao for being fantastic mentors; Steve Einbinder for showing me the power of LabVIEW; Arun Kumar and Hua Li for being great technical resources, sounding boards, and friends; Fred Harris, Bernard Sklar, and Jim McClellan for being the three instructors who I learned the most DSP from; and last but not least, DP for all his creative inputs and edits.

Finally, I would like to say that I have made every effort to put only correct information in this book, but I am sure that I made at least a few mistakes. If you find an error, I would be happy to hear about it. Also, please send any comments or suggestions to me at *coryc85@gmail.com*. Thanks and enjoy the book.

Cory L. Clark

LabVIEW
Digital Signal
Processing

1

Getting Started

Digital Communications and LabVIEW

When most people think of digital communications, they probably imagine that it has something to do with computers talking back and forth. However, what they do not realize is that digital communications is really about sending digital data through some kind of medium that is much more suited to analog signals. Let's face it—the world is an analog place and, in fact, digital communications actually involves the transmission of a set of discrete analog waveforms. Perhaps digital communications should be renamed discrete communications. At least we can say that each discrete waveform has an associated digital representation. So in this book a digital communication system means any system where digital data are transmitted from one place to another using some finite signal set. Digital communication systems are everywhere. Cellular telephones, hard disk drives, DSL modems, satellite television, CD players, and even your garage door remote are all examples of a communication system where digital data are transmitted. For the many different communication applications that exist, there are even more digital communication protocols: GSM, CDMA, OFDM, 802.11(b and g), Ethernet, APCO-25, as well as emerging protocols such as EDGE and W-CDMA.

The list of wireless digital communication standards in Table 1.1 is only a fraction of the digital communication systems in use today and each of them has its own unique frequency band, signaling format, and multiple access method. These systems are becoming so ubiquitous that every company involved in communications has to sell a product that supports Bluetooth, 802.11, Infrared, IEEE-1394, USB, and more. It seems as if there is a convergence of communication devices but a divergence of communication protocols. Test and development systems to cover all of these standards are expensive and inflexible—it is difficult to buy equipment that can keep up. So what can we do to keep up with the rapidly evolving technologies in digital communications? Better simulation and prototyping might be one solution, but how do you really test the whole system?

TABLE 1.1   **Wireless Digital Communication Standards**

Communication protocol	Frequency band(s)	Channel bandwidth	Modulation
GSM	800/900 MHz	200 kHz	GMSK
CDMA	800/900 MHz and 1.9 GHz	1.25 MHz	QPSK
Digital cellular	800/PCS	30 kHz	π/4-DQPSK
TETRA	400/800 MHz	25 kHz	π/4-QPSK
802.11b	2.4 GHz	22 MHz	CCK
802.11g	2.4 GHz	22 MHz	OFDM
Bluetooth	2.4 GHz	1 MHz	GFSK
APCO-25	VHF, UHF, 800 MHz	12.5/25 kHz	C4FM
GPS	1.2 GHz and 1.5 GHz	20 MHz	Binary biphase
CDMA2000	1.9 GHz	1.25 MHz	4-PSK
UMTS	900 MHz and 1.9 GHz	1.25 MHz	GMSK/8-PSK
EDGE	900 MHz and 1.9 GHz	200 kHz	GMSK/8-PSK

Many engineers turn to MATLAB or perhaps even to C to simulate their communication designs before starting production. Both are excellent tools and are extremely powerful. However, for the whole package, hardware to bits, LabVIEW really outshines both of them. From a general point of view, LabVIEW's graphical environment is easy to pick up and understand. Looking a little deeper you can see that LabVIEW has a powerful built-in signal processing toolset, no coding requirements such as memory allocation or variable declarations, no compiling, highly integrated instrument control or data acquisition, and excellent display utilities for viewing these digital signals at various points in the communication system. We will see how simple LabVIEW virtual instruments (VIs) can be built and combined to produce a flexible and powerful digital communication test system. Building such a LabVIEW-based communication system will allow for new interface standards to be quickly and easily integrated. Let me also say that LabVIEW has come a long way as far as speed in the last several years is concerned, but it is by no means a real-time environment. As we develop the pieces of our digital communication system it is important to realize that this system is really too slow for real-time communications, but is perfect for instrument-type measurements. Building this system in LabVIEW is a trade-off in speed for rewards in flexibility, ease of coding, and excellent user interface and data display.

LabVIEW has always been one of the premiere tools for instrument control and data collection or display and this book will show how LabVIEW can be adapted to process any of these digitally modulated signals—effectively becoming the test instrument itself. We will see that the line between instrument and virtual instrument (VI) can be blurred. To accomplish this we will examine the pieces of a digital communication system in a LabVIEW environment and assemble those pieces into a digital receiver. Of course the LabVIEW tools that are developed in this book are by no means considered real time. There will be a lot of overhead due to the PC running Windows, then overhead from LabVIEW— all on top of the computations and display updates involved in demodulating our signals. So we may sacrifice the speed of a stand-alone dedicated instrument to gain the flexibility of demodulating any signal we choose. The point is that this digital communication system is really more of a testbed than a radio. We will be able to simulate, test, and measure the critical aspects of a real communication system. Throughout the book, some ways to optimize the speed of LabVIEW's processing will be mentioned and Chap. 9 will offer some helpful tips for maximizing throughput.

Let us start by exploring exactly how LabVIEW can be used for receiving digital communication signals. This book will focus on the implementation of two different digital communication structures. The first is a typical digital receiver, discussed in Sec. 1.1. Section 1.2 covers the second type, which is an all-digital receiver. Both receivers are used in practice today and both have their positive and negative qualities. The following sections will outline the advantages and disadvantages of each type of digital receiver and give the reader a better idea of how exactly LabVIEW fits into these structures.

Before getting into the specifics, let us review how digital frequency relates to analog frequency. Digital frequency will be denoted by Ω, where

$$\Omega = 2\pi \frac{f}{f_S} \tag{1.1}$$

Since digital samples themselves are nothing but a sequence of numbers, they possess no inherent time information. Therefore, by specifying the sample period (the time between successive samples), we can relate the analog frequency f to the digital frequency Ω, where the unit of Ω is radian per sample. Therefore, in the following figures Ω will imply any processing that is done in the digital domain and f will imply any processing that is done in the analog domain.

1.1 Conventional Digital Receiver

Figure 1.1 is the block diagram of a typical "digital" receiver. In reality this receiver is only partially digital because there is analog processing that takes place at the front end of the receiver. Before the desired signal can make it to the digital world for processing by the DSP, it must first be sampled by the analog-to-digital (A/D) converter. A standard A/D can sample up to 20 megasamples per second (Msps), which means it probably has a front-end filter (not shown) limit of 10 MHz. Unless the desired radio frequency (RF) signal is below 10 MHz, the A/D will not pass the signal through to the digital processing section. It is for this reason that heterodyne is necessary. The incoming RF signal is first mixed with a local oscillator (LO) signal to shift the carrier frequency from some high range down to an intermediate frequency (IF) range more suitable for the limits of the A/D. This IF signal is then passed through an analog IF filter before digitization to remove any unwanted signal energy and thereby improve the sensitivity of the receiver.

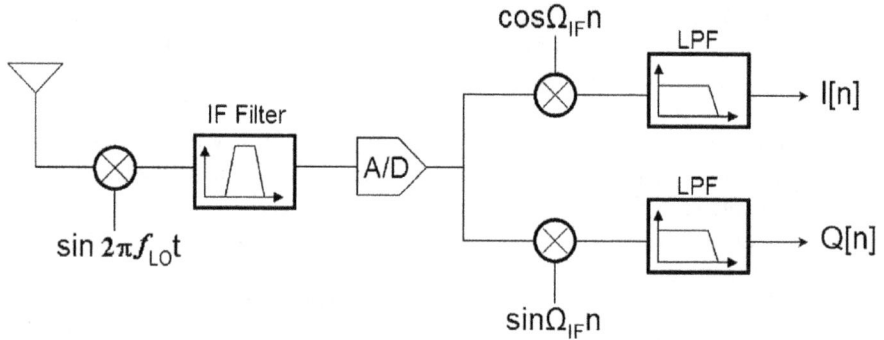

Figure 1.1 Conventional digital receiver.

Assuming that the mixer has a method to reject the images, this type of receiver will preserve the signal-to-noise ratio (SNR) of the original signal [1]. This property may be necessary for applications where the SNR is low to begin with.

The conventional digital receiver described previously, although widely used, becomes cumbersome to implement in a software environment such as LabVIEW. The downconversion from RF must be done in hardware with an analog mixer. The LO signal is typically generated by some voltage controlled oscillator circuitry and amplified to provide the appropriate drive level for the mixer. This requires external hardware in addition to the sampling card. Also, any impurities in the LO signal immediately affect the quality of the mixer output. Additionally, LO stability and its tuning range place hard limits on the signal bandwidth and carrier frequencies that can be analyzed with this type of receiver. However, these types of receivers are available. In fact, National Instruments, as well as other companies, offer RF downconverter products that provide a complete black box solution to LO generation, RF mixing, and IF filtering. A few of these products and vendors are listed in App. B. Chapter 2 will take an in-depth look at the National Instruments PXI-5660 RF Signal Analyzer. It is worth noting that the PXI-5660 is in some sense a hybrid device in that aspects of both the conventional digital receiver and the subsampling receiver are used to acquire the desired signal. The details of the PXI-5660 operation will be explained later in more detail; however, from the casual user's point of view the 5660 device appears to be a conventional digital receiver and this book will treat it as such.

1.2 Subsampling Receiver

While the conventional digital receiver is perfectly acceptable for performing each and every function in this book, we can modify the structure in Fig. 1.1 to eliminate much of the hardware. This is done by moving the A/D over to the

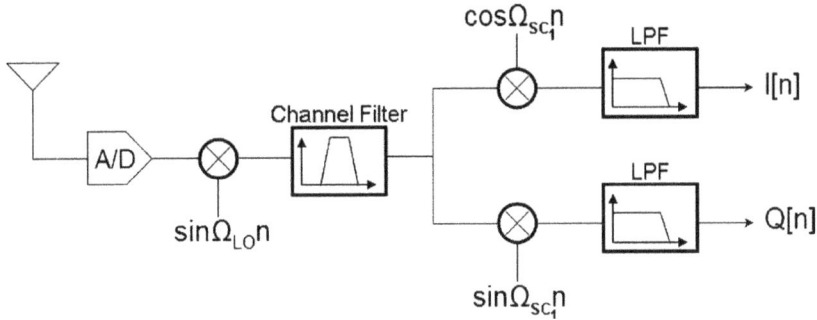

Figure 1.2 All-digital receiver.

high-frequency side of the LO mixer and sampling the RF directly, creating an all-digital receiver. Figure 1.2 shows an example structure of what this all-digital receiver might look like. In this type of system, the A/D must now sample the RF signal directly and the LO has become a simple digital mixer. All of this implies that the A/D has the bandwidth and speed necessary to capture the desired signal. As before, if the RF carrier frequency is low enough, standard A/D converters can easily capture the signal, but many modern communication systems operate with spectrum in the 800 MHz to 1 GHz range. Without violating the Nyquist rate, the A/D must be able to sample at twice the largest input frequency, potentially over 2 gigasamples per second (Gsps). While that is a staggering sampling requirement, there are A/Ds available with plenty of bandwidth and sample rates up in the gigasamples per second. In fact, Acqiris makes a digitizer that has 1 GHz of analog bandwidth and captures up to 2 Gsps. Chapter 2 will make use of this particular device to capture a digital communication signal.

One headache that always arises when we start sampling signals in the gigahertz range is the huge amount of memory space required to store those samples. Assuming the samples are 8-bit real samples and the sample rate is 1 Gsps, we can fill a gigabyte of memory space up with RF samples just capturing a 1 s time record. This does not leave much room in the memory for the operating system or any other applications. On top of that, any processing on that enormous number of samples requires long periods of time, which is an undesirable result for any pseudo real-time testing. So how can we simultaneously take advantage of the large analog bandwidth of some of these A/D converters without incurring the costs associated with processing billions of samples per second? One way is to deliberately violate the Nyquist sampling theorem and sample at a rate much less than twice the highest frequency component in our signal. This technique is known as subsampling, bandpass sampling, or undersampling and there are tight limits on the range of sample rates that will produce the desired result without distorting the spectral replications.

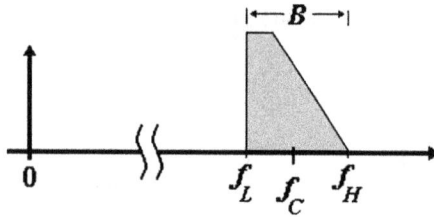

Figure 1.3 Spectrum of real signal in analog world.

Let us start by examining the spectrum of a real signal shown in Fig. 1.3. For descriptive purposes only the positive half of the spectrum is displayed. This signal has a bandwidth B that extends from f_L to f_H. We can see from the picture that the signal is centered at an arbitrary carrier frequency f_C. Clearly,

$$f_L = f_C - \frac{B}{2} \tag{1.2a}$$

$$f_H = f_C + \frac{B}{2} \tag{1.2b}$$

The conventional rule of thumb for sampling the analog signal in Fig. 1.3 would be to sample at a rate greater than twice the highest frequency contained in the signal f_H. This is exactly the scenario shown in Fig. 1.4a. Here both positive and negative frequencies as well as the first spectral replications due to the periodic sampling are shown. We can see that sampling at some f_S greater than twice f_H will leave a guard band between the spectrum of the sampled signal and the half sample rate π. The sample rate drops to exactly $2f_H$ in Fig. 1.4b and the spectral replications slide closer to the half-sample rate or away from

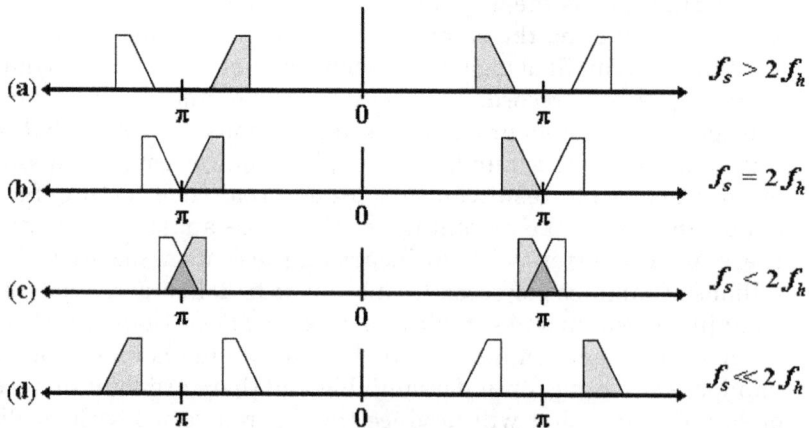

Figure 1.4 Spectral translation.

multiples of 2π, leaving no guard band at all. In Fig. 1.4c, the sample rate has now crossed a threshold and is actually violating the Nyquist criterion by sampling below f_H. At this point, there is a range of sample rates that will result in undesired aliasing; however Fig. 1.4d has continued the reduction in the sample rate and f_S is now safely below f_H with no aliasing. Interestingly, from the figure we can see that the spectral replications have now swapped places and are inverted in frequency. In fact, in [2] Liu shows that the orientation of the spectral replications switches from normal to inverted at each integer reduction of the sample rate. As the sample rate drops, the spectral replications continue to march away from multiples of 2π and are inverted at times. The range of acceptable rates at which we can sample our signal without aliasing is given by [3]:

$$\frac{2f_H}{n} \leq f_S \leq \frac{2f_L}{n-1} \tag{1.3}$$

n will be denoted as the subsampling factor and is given by [3]:

$$1 \leq n \leq I_g\left[\frac{f_H}{B}\right] \tag{1.4}$$

where $I_g[\frac{f_H}{B}]$ is the largest integer contained in the argument.

From Eq. (1.3) it is clear that there are many possibilities for choosing how much to undersample the desired signal. One obvious choice would be to sample at the absolute lowest possible rate, or said another way, choose the largest possible n. Substituting Eq. (1.2b) in Eq. (1.3), we can rearrange the lower limit in Eq. (1.3) to show the absolute minimum sample rate as

$$f_S \geq \frac{2f_C + B}{n} \tag{1.5a}$$

using Eq. (1.4), we know that the largest value for n will always be $[\frac{f_H}{B}]$ and again substituting for f_H using Eq. (1.2b) we will get the following:

$$f_S \geq \frac{2f_C + B}{\left(\frac{f_C + B/2}{B}\right)} \tag{1.5b}$$

now we can rearrange the denominator by factoring out the 1/2B term

$$f_S \geq \frac{2f_C + B}{1/2B(2f_C + B)} \tag{1.5c}$$

finally, we can move the $2B$ term to the numerator and cancel the $(2f_C + B)$ terms leaving only

$$f_S \geq 2B \tag{1.6}$$

which tells us that the *absolute* minimum rate at which we can sample a signal is twice the information bandwidth B.

Why would I not just sample at the lowest possible rate given in Eq. (1.6) as $2B$? Well there are several considerations for choosing the subsampling factor n. First of all, it is important to understand the effect of odd and even factored reductions. As mentioned earlier from [2], if n is even the resultant signal spectrum will be inverted and the spectrum will be normal if n is odd. Depending on the application, one or the other orientation may be desired. In the case where the subsampled signal ends up in the wrong orientation, the spectral inversion should be easy to fix but may need some additional processing steps. Secondly, the choice of n will affect the frequency where your aliased image will end up. Akos et al. in [1] shows that the IF frequency can be computed from

$$\text{fix}\left(\frac{f_C}{f_S/2}\right)\begin{cases} \text{even,} & f_{IF} = \text{rem}(f_C, f_S) \\ \text{odd,} & f_{IF} = f_S - \text{rem}(f_C, f_S) \end{cases} \tag{1.7}$$

where $\text{fix}\left(\frac{f_C}{f_S/2}\right)$ is simply the integer portion of the argument.

This means that you as the designer will have a measure of control over where to put the aliased spectra. The third and possibly the most influential consideration for choosing n is the resultant degradation in SNR produced by the subsampling. As with almost everything in engineering, we do not get something for nothing. In this type of sampling, the trade-off is in SNR [3]. By undersampling, we also alias noise into the translated passband of our signal; this reduction in SNR is unavoidable, but can be acceptable for the conducted signals that we are going to encounter in this book. In [3], the degradation in SNR D is given by

$$D \approx 10 \log n \tag{1.8}$$

From Eq. (1.8), it is evident that as we increase n (or decrease the sample rate) we are folding more and more noise into the passband of our signal. In Chap. 2, when we start using some hardware we will actually take a look at a real-world signal sampled both at a rate above the Nyquist rate and undersampled to a rate consistent with Eq. (1.2). We will be able to clearly see the implications of the undersampling on our signal. The overall effect of the signal degradation D will depend on the required SNR for the specific communication system.

With all these restrictions on valid sampling rates and SNR reduction, subsampling seems much more complicated than the plain old Nyquist rate sampling method; however, this approach has two wonderful results. First, we can enormously reduce the required sample rate and therefore use much less memory space to capture and process the same RF signal. Second, if we choose our sample rate appropriately, our aliased signal will end up right at the IF of our choice and we will have completely eliminated the need for the digital LO mixer operation shown in Fig. 1.2. The only question will be whether or not we can accept the reduced SNR without a severe impact to our recovered signal.

Example 1.1: Subsampling a GSM Signal   Assume that you are trying to sample a GSM cellular signal. The carrier frequency is 1 GHz and the signal bandwidth is

30 kHz. If we obey Nyquist, we must sample above 2 Gsps, but with subsampling, we can use Eq. (1.3) to choose a much lower sample rate. The sample rate floor is still determined by Nyquist; remember that f_S must always be at least twice our signal's information bandwidth, or in this case 60 kHz. The GSM standard calls for 9-ms slots with a call interleave of 3:1. If we want to sample 10 occupied slots, the capture length would have to be 270 ms. At 2 Gsps, this amounts to 540 million samples, but using subsampling, our sample rate and the number of samples can be reduced according to Eq. (1.4) as

$$1 \le n \le I_g \left[\frac{1\,\text{GHz} + 15\,\text{kHz}}{30\,\text{kHz}} \right] \quad \text{or} \quad 1 \le n \le 33{,}333 \qquad \text{(a)}$$

If we choose $n = 10{,}000$ that means we can sample the SAME 270-ms time signal in only 54,000 samples!

Summary

LabVIEW provides an ideal environment for simulating and testing digital communication systems for several reasons. First of all, its graphical nature allows the engineer to quickly test components without all of the overhead found in typical code or compiler systems. Second, LabVIEW was conceived to interact with physical instruments and thus the acquisition of real signals is typically straightforward and efficient. And third, we will see that LabVIEW has loads of built-in signal processing tools that are simple to drop into a VI and start using. And finally, you will find that most of the hardware available out there is compatible with LabVIEW.

We have seen that LabVIEW can accommodate both a conventional digital receiver and an all-digital receiver. Both receivers are capable of analyzing every digital signal presented in this book. However, each receiver has unique subtleties that may or may not be important to your particular application. We have also discussed the usefulness of subsampling receivers. Because of the limitations on the speed of processing millions (or billions) of samples and memory requirements, subsampling is an attractive way to acquire a digital communication signal. There are some very specific restrictions on valid sample rates for these types of receivers and there is also a cost associated with undersampling. We will see more on this subject later on, but first we will jump to building our digital receiver in Chap. 2 by outlining various hardware devices for acquiring the desired signal. Chapter 3 will focus on LabVIEW's spectral analysis capabilities. The concepts and tools developed there will be used later in Chap. 4 while analyzing digital filters. LabVIEW has an impressive selection of built-in filtering routines and we will build some complete filter design tools around those routines. Multirate processing will be covered in Chap. 5 and Chap. 6 will get us started generating some useful signals such as mixers and noise. Then Chaps. 7 and 8 will start putting all of this together into a complete communication system. Finally, Chap. 9 will finish up with some tips to optimize the speed of your LabVIEW processing.

References

1. Akos, D. M., M. Stockmaster, J. B. Y. Tsui, and J. Caschera, "Direct Bandpass Sampling of Multiple Distinct RF Signals," *IEEE Transactions on Communications,* vol. 47, pp. 983–988, July 1999.
2. Liu, J., X. Zhou, and Y. Peng, "Spectral Arrangement and Other Topics in First-Order Bandpass Sampling Theory," *IEEE Transactions on Signal Processing,* vol. 49, pp. 1260–1263, June 2001.
3. Vaughan, R. G., N. L. Scott, and D. R. White, "The Theory of Bandpass Sampling," *IEEE Transactions on Signal Processing,* vol. 39, pp. 1973–1984, September 1991.

Getting a Signal into LabVIEW

Before we can start performing any sort of digital processing with LabVIEW, we have to somehow obtain the signal that we would like to analyze. Chapter 1 outlined two types of digital receivers. The first type was the conventional digital receiver that heterodynes the radio frequency (RF) signal down to a frequency suitable for most analog-to-digital (A/D) converters to capture. The second type was the subsampling receiver that samples the RF signal directly. As mentioned before, both receivers are capable of processing the signals that we explore in this book and both have their pros and cons. This chapter focuses on the implementation of each of these receiver structures. In later chapters we will see some methods for testing the receiver using our own test signals within LabVIEW. As outlined in Chap. 1, LabVIEW is particularly well suited for interfacing to the physical world. For instance, the interface might be through dedicated instruments over a general purpose interface bus (GPIB) connection. As we develop the structure of the receiver, we will see that we can use an unlimited number of input hardware devices to actually sample a signal and the subsequent signal processing does not change at all. The virtual instruments (VIs) shown here will be used as building blocks to create a complete digital communication system. All VIs are referred to by name and are available for download from the website *http://www.MHEngineeringResources.com*. Also Appendix A contains a complete listing of each VI along with a description and block diagram.

2.1 Conventional Digital Receiver

If you have ever looked at the design of a digital receiver before, it probably looks a lot like Fig. 1.1. In a band-specific RF product, typically the local oscillator (LO) mixer, intermediate frequency (IF) filter, and A/D converters are combined into a single receiver front-end IC with RF in and baseband I and Q digital samples out. This is a very nice solution for a dedicated receiver, but for our flexible LabVIEW receiver, it might get a little tricky. To quickly implement this receiver, we need a black box that takes in a high-frequency RF signal with some specified bandwidth and pipes digital samples right into LabVIEW. You could also build your own RF interface to perform the filtering and mixing and present a suitable IF signal to some off-the-shelf A/D acquisition card. But the focus of this book is really on the digital communication capabilities of LabVIEW and not on the design of RF hardware. National Instruments does make some modular equipment to perform the functions that we need for this receiver. Specifically, we can use the PXI-5600 RF down-converter and the PXI-5620 IF digitizer. NI packages these two devices together into the PXI-5660 RF signal analyzer. Before we get too far into details, I want to give a brief overview of the PXI family of devices. PXI is the National Instruments version of compact PCI. The instruments are modular and plug into slots in the PXI chassis. The chassis itself needs either a self-contained controller, sold as a module, or a kit that allows a stand-alone PC to act as the controller. NI calls the interface to an external PC their MXI solution and it requires an expensive copper or fiber optic cable for communication between the PC and PXI chassis instruments.

Figure 2.1 shows a typical PXI configuration. At a minimum, the PXI chassis needs a controller (either embedded PC or the MXI PC interface), a power supply, and an instrument for some type of measurement or acquisition. For building a conventional digital receiver on this platform, the PXI-5600 RF down-converter and the PXI-5620 IF digitizer along with the PXI chassis itself and its power supply will be absolutely necessary items. Your choice of controller will

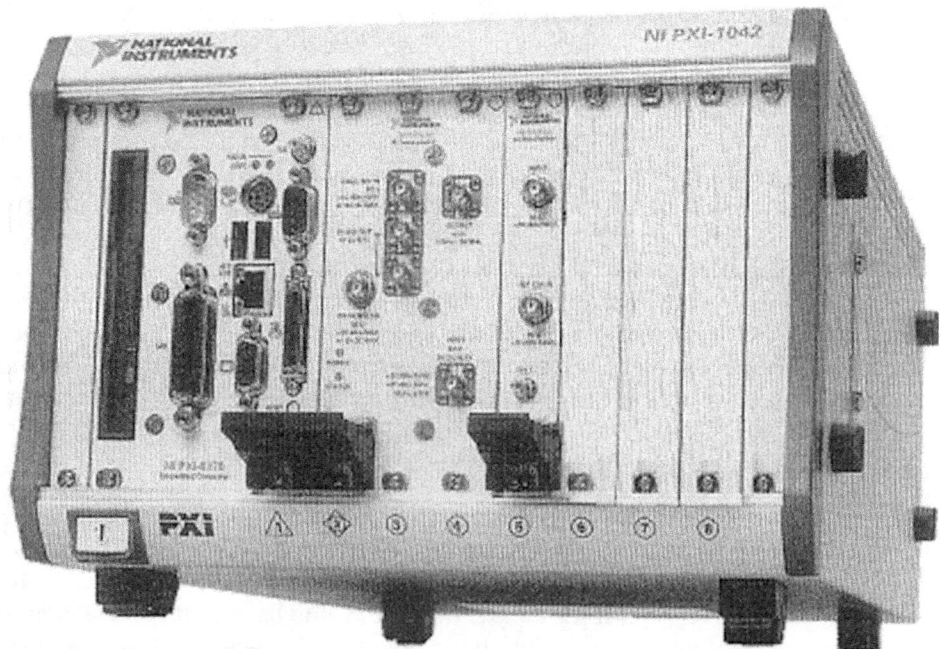

Figure 2.1 PXI chassis with modules.

depend on how much you want to spend and the specifics of your application. The MXI interface to an external PC will be the cheaper route assuming that you have already got the PC (approximately $1500 cheaper), but will also be the bulkier route. A National Instruments embedded controller will cost about $4000, but will also be a more compact solution. One other crucial piece of information is that the NI embedded controllers are typically based on laptop computer technology and are not always as powerful as comparable desktop machines in the market. The top-of-the-line embedded controller, the 8186, currently has a 2.2 GHz processor with up to 1 GB of RAM while a top-of-the-line desktop PC has a 3.4 GHz processor and can handle up to 4 GB of RAM. The performance difference will be noticeable in how fast your signal samples are processed and in any updates to the displayed data. Please note that NI does not ship a monitor, keyboard, or a mouse with their embedded PCs and there is no CD drive and no floppy (although the CD drive is an option). USB or Ethernet are the best available methods for loading files onto these controllers.

Now that we have explored the various options for building our conventional digital receiver, let us look at getting a working setup. Since the controller is transparent to the PXI hardware, we can ignore the actual controller interface chosen for your specific application. What is common, however, is the LabVIEW interface to the PXI instruments. In order to use any of the PXI modular instruments, we will need (in addition to LabVIEW) NI Tuner and NI Scope. These two pieces of software typically ship with all PXI modules and are also downloadable from the NI website. National Instruments also bundles the Spectral Measurements Toolset with the 5660 package. This toolset has some nice zoom FFT functions, spectrograms, and most importantly offers some very good examples for developing your own communications-related programs. Chapter 3 will explore the SMT package in more detail. Once you get all of these items installed and you reboot, your function palette in LabVIEW should look like Fig. 2.2.

So what are all these new functions? Well, the digitizer pane will contain all the functions necessary to initialize and use the PXI-5620 digitizer. National Instruments calls its standard interface to all digitizers NI Scope, which all the functions here refer to. The radio frequency signal analyzer (RFSA) pane contains all the functions associated with the PXI-5600 downconverter (generically called NI Tuner) and RFSA also contains some special functions with names that begin with SMT. These functions are specific combinations of the NI Scope and NI Tuner functions designed especially to set up the PXI-5660 hardware for capturing RF signals. Everything that is done inside the SMT functions is available to you from the digitizer and RFSA subpalettes shown on the right in Fig. 2.2. For inexperienced LabVIEW users or anyone new to the PXI platform, the best route for getting the 5660 hardware working will be to rely on the SMT functions as they provide for fairly quick and effortless configuration. Later as you gain skill with the 5660, you can always go back and modify your VIs to use the lower level calls to the hardware. You will probably find later that your application really does not require all the bells and whistles provided in the examples and you can easily tailor the example files to suit your specific needs.

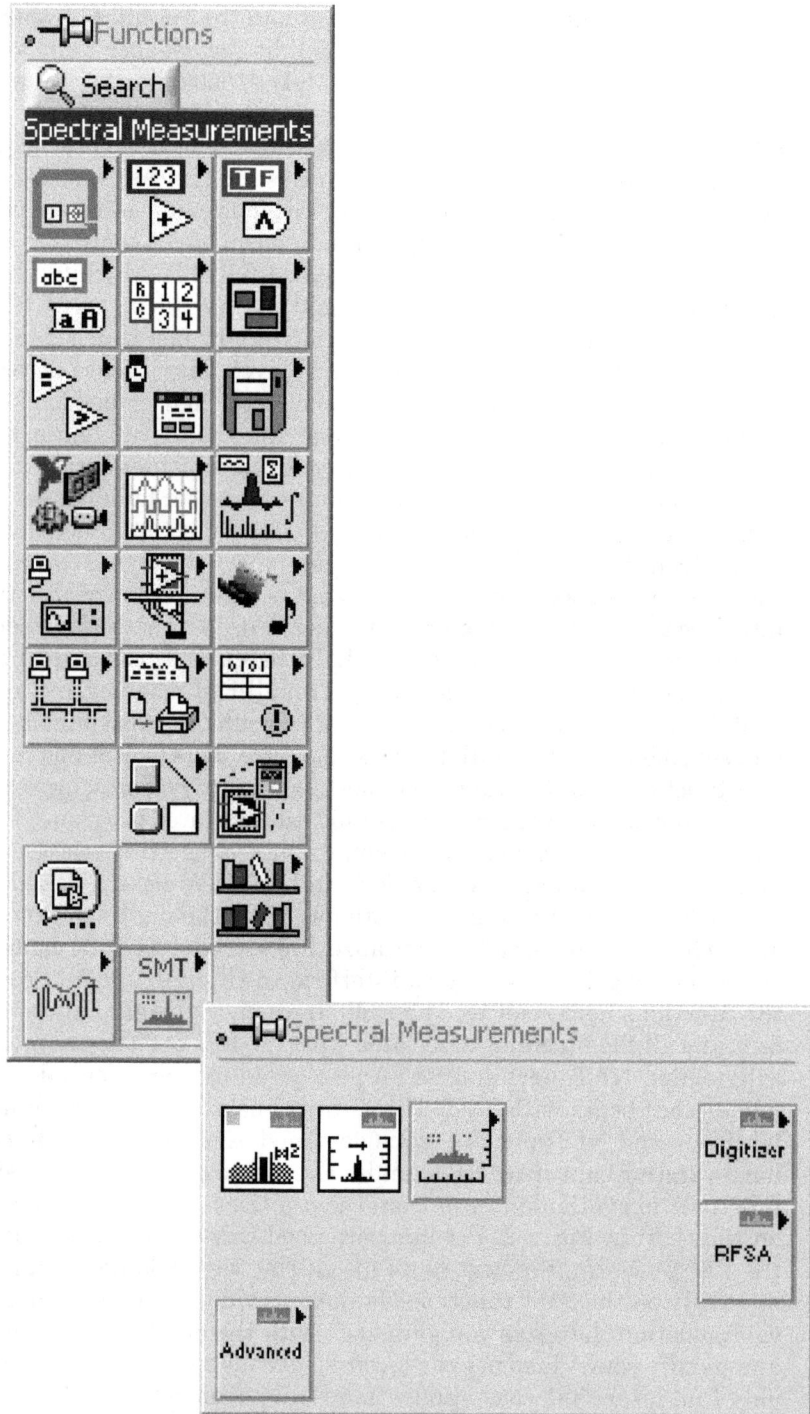

Figure 2.2 Spectral measurements function palette.

The simple VI shown in Fig. 2.3 is a very basic program to capture data from the 5660 RF signal analyzer. This file is called PXI Capture.vi and works well for a quick solution to acquire and display a signal. Like most LabVIEW compatible instrument drivers, the PXI functions can be categorized into three groups: initialization, acquisition, and resource deallocation. The two sub-VIs on the left are Init5660.vi and Config5620.vi. These functions fall into the category of initialization. Those two functions are simply reconfigurations of the standard NI SMT examples to separate the initialization routine from the acquisition. Opening resources in LabVIEW is much the same as opening files in C. The resource is initialized only once (like fopen in C) and the initialization returns a handle or an ID to that resource. The handle or ID *must* be used in any subsequent calls to access that instrument. This is analogous to a file pointer in C. Inside the loop, we continually read from that resource and once the loop terminates, we close the resource. Notice that the PXI instrument parameters do not allow the user to choose the sample rate. In a minute we will see why National Instruments decided to keep control over the sample rate. For now we can live with the fact that we always inherently *know* the sample rate because you see that one user input is the time length of the capture and we can use an array size function to calculate the number of points captured. Hence the sample rate is the number of samples divided by time in seconds. Later, in Chap. 5, we will discuss multirate signal processing and build some tools to actually resample the data to another sample rate. Building a resampler means that we can work with whatever rate the hardware gives us, but at the expense of adding more computations to our receiver.

So why does NI control the sample rate? The National Instruments product information for the PXI-5660 promotes the accelerated throughput time of the 5660 device versus standard instruments. In order to speed up that availability of samples from the hardware, NI has come up with an optional sampling scheme called direct digital conversion or DDC. Without knowing specifically what is going on inside the 5660 package, we can see from the help files that the samples are output at a rate that is related to the information bandwidth of the signal and not related to the carrier frequency. In fact, the help file for one of the NI Modulation toolkit VIs shows the following information.

Going back to the discussion in Chap. 1 about subsampling a signal, it is clear that the NI hardware is performing the same operations that Eqs. (1.2) and (1.3) describe. In other words, the sample rate can be reduced and the sampled signal can be directly translated to an IF via the sampling process much like what we will show in Sec. 2.2 with our own sampling hardware. Thus, Table 2.1 shows that NI allows you to have *limited* control over the use of the DDC feature by setting the desired signal bandwidth of the capture.

For further information on the VIs shown previously, please refer to Appendix A. Here a complete reference listing of all VIs used in this book is provided along with a brief description of their function. Probably the best place to start for anyone new to the PXI world is with the examples provided by National Instruments. For the PXI-5660, the examples are typically in the folder C:\Program Files\National Instruments\LabVIEW 7.0\examples\Spectral

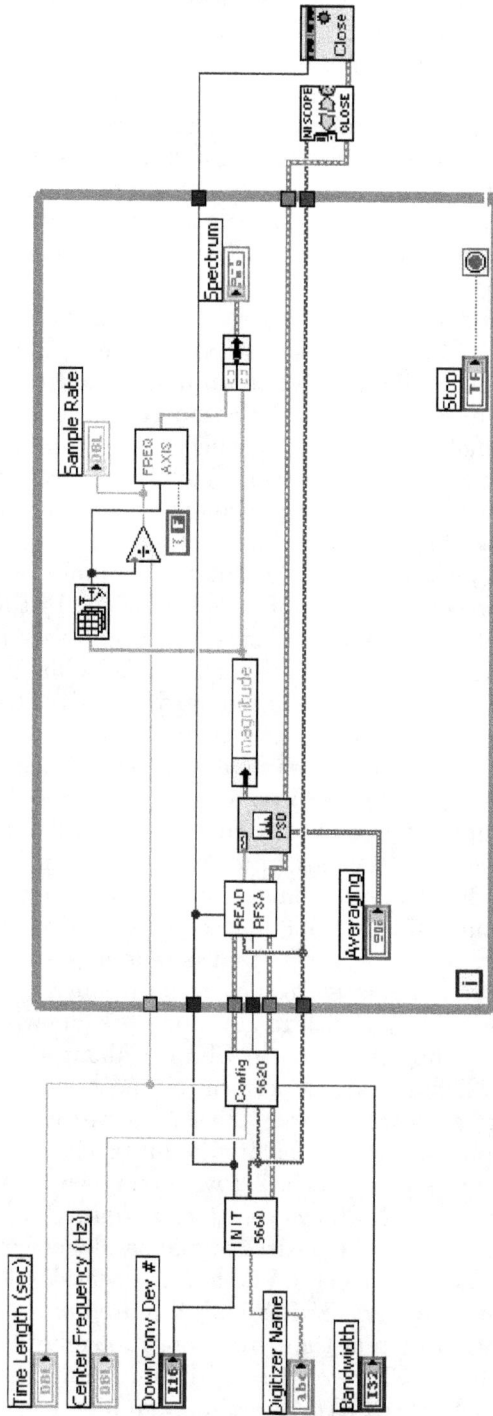

Figure 2.3 Simple PXI-5660 capture using PXI capture.vi.

TABLE 2.1   DDC Signal Bandwidth versus Sample Rate

Signal bandwidth	Highest available sample rate
1.25 MHz	2 Msps
800 kHz	1 Msps
400 kHz	500 ksps
200 kHz	250 ksps
100 kHz	125 ksps
50 kHz	62,500 sps
25 kHz	31,250 sps
12.5 kHz	15,625 sps

Msps = megasamples per second; ksps = kilosamples per second; sps = samples per second.

Measurements Toolset\SMT Examples for RFSA. Browsing around in this folder, you should find some very useful programs to get you up and running with the hardware. Many of these examples are made specifically to be generic in their end use, so be aware that there will be many parts of these VIs that are completely unrelated to your application. Also remember to make use of the LabVIEW help files for any functions that are unfamiliar.

2.2   Subsampling Digital Receiver

Chapter 1 referred to the subsampling receiver as an all-digital receiver. That terminology stems from the fact that the only piece of hardware in the entire system is the A/D converter and *all* of the signal processing is done digitally. Because of this, one of the most important properties of the A/D converter we choose will be its analog bandwidth. If the A/D converter has any front end filters that limit the input bandwidth to below our carrier frequency, then we can never get the signal through the sampler. You might have been thinking that the sample rate is the most important characteristic, but as discussed in Sec. 1.2, we can choose almost any arbitrary sample rate down to $2B$ and still have an accurate digital representation of our signal. So, when choosing a sampling card, look first for one with enough analog bandwidth to squeeze your signal through the front end and then find one that can handle your range of sample rates.

Surprisingly most A/D cards have bandwidths only up to 10 or 20 MHz. This is an acceptable limitation if your signal's carrier frequency is low enough; however, the majority of wireless digital communication is done up in the 800 MHz to 1 GHz range. That makes it tougher to find a suitable card but certainly not impossible. To make your search a little easier, Appendix B is a list of sampling hardware equipment manufacturers and a brief overview of what they have to offer. I found that Acqiris (*www.acqiris.com*) offers a line of PCI-based 8-bit sample cards with bandwidths from 150 MHz up to 1 GHz that provide precisely what we need for building this subsampling digital receiver. All of these cards are under $10,000 and the best part is they provide built-in LabVIEW functions for interfacing to their hardware. The rest of this section focuses on using the

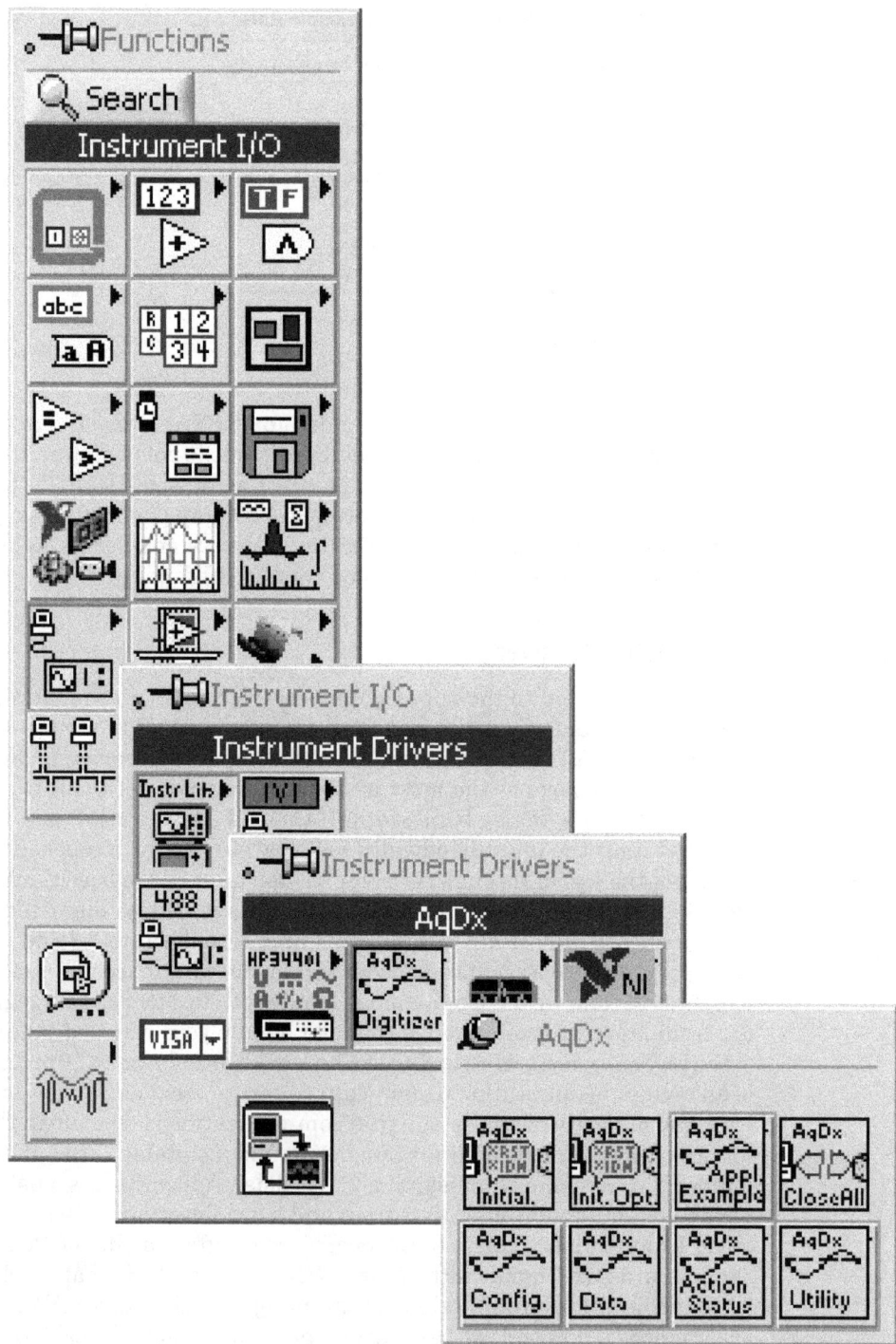

Figure 2.4 Acqiris digitizer functions palette.

DP240 8-bit PCI sampling card. This device has two input channels, 1 GHz of analog bandwidth, and can acquire up to 2 gigasamples per second (Gsps) (shared over both channels). The card also comes with the necessary drivers for LabVIEW on a CD. The installation is simple and when it is finished, the LabVIEW functions palette looks like Fig. 2.4.

In addition to the basic driver functions Acqiris also provides an example VI to get you started using their card quickly. The example is called Acquire.vi and is a simple and straightforward program. The VI block diagram in Fig. 2.5 shows a slightly modified version of the standard Acqiris example. In this VI, the basic Acqiris functions have been separated into the three groups mentioned previously and the acquisition subVI is placed inside a loop for continuous acquisition of the signal.

By incorporating a power spectral density VI from the signal measurements palette inside the Acqiris samples sub-VI, the spectrum of our signal can be computed for each time record captured and displayed on the front panel. In addition, the frequency axis sub-VI is added to generate the x-axis frequency values for a given sample rate. These two functions can then be used to create a display such as the one shown in Fig. 2.6. The main point here is that with some very simple modifications, the standard example program can be modified to include whatever functionality that we need in our application. The VIs shown in this section will be used later as we build the other pieces of our digital receiver.

Let us take a closer look at Fig. 2.6 before moving on. The resource name (PCI::INSTR0) is specific to your device. In this case, the Acqiris card is PCI device 0. If you are unsure of your hardware setup, you can use the National Instruments Measurement & Automation Explorer (MAX) to view all installed NI compatible hardware. The figure shows controls for the desired number of samples and the sample rate. Just how do you determine the sample rate? Chapter 1 gave some guidelines for appropriate subsampling rates and the impact of those rates on the captured signal. The next three sections explore those topics in more detail.

2.2.1 Choosing a sample rate

In contrast to the National Instruments PXI interface, the Acqiris digitizer functions allow the user to specify the sample rate of the acquisition. This ability is of utmost importance when building a subsampling receiver. The only concern for the user is what sample rate to choose. The lowest usable sampling rate is given by Eq. (1.5) as $2B$, but we can also choose any arbitrary rate up to the limit of our sample card, as long as the chosen sample rate does not fall into one of the forbidden zones where the aliased images overlap. These zones are defined by the Eqs. (1.3) and (1.4) and are depicted graphically in Ref. [1]. Depending on the processing that is done after the signal is digitized, there may be a need to sample the data at a particular rate. Outside of those specific processing requirements, why don't we just go right to the absolute lowest rate possible? There are really two major considerations for choosing a good subsampling rate:

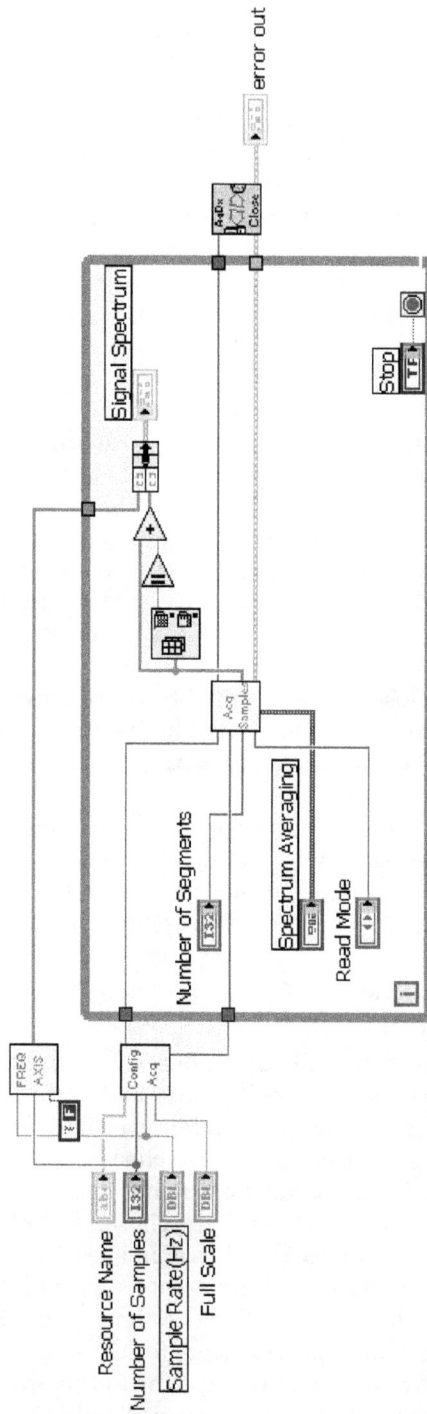

Figure 2.5 Modified Acqiris example.

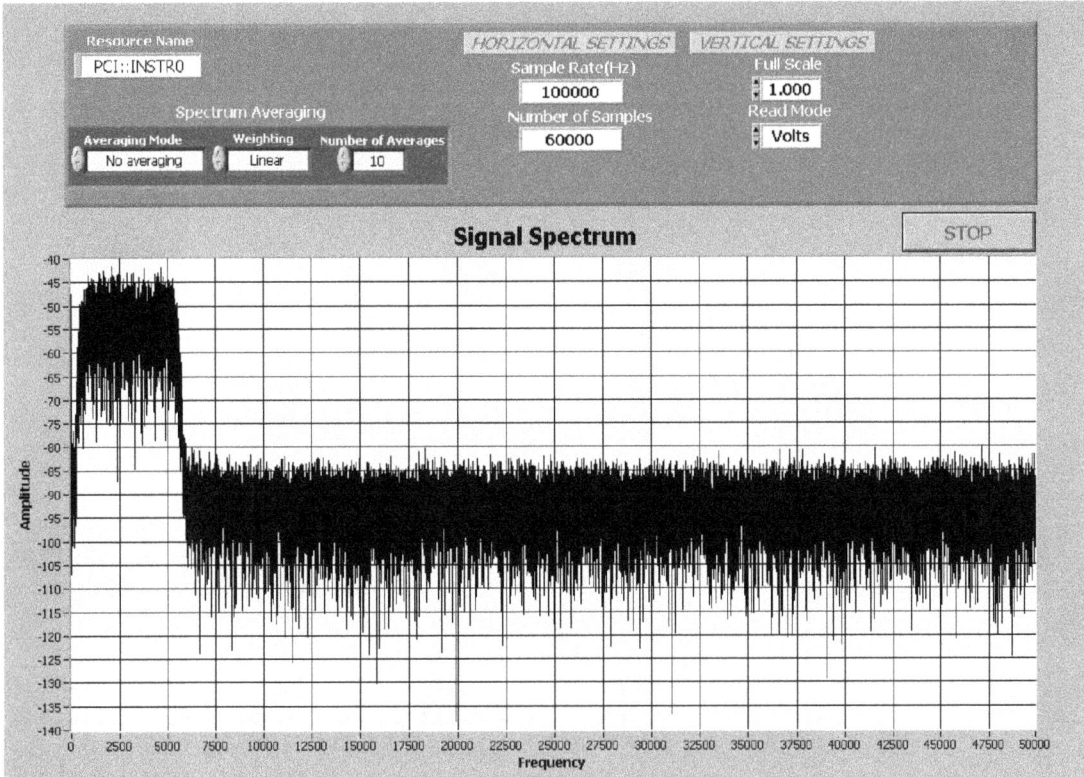

Figure 2.6   Front panel for Acqiris acquire.

signal-to-noise ratio (SNR) degradation and subsampled signal placement. Chapter 1 presented the theoretical implications of subsampling both in the SNR that is shown in Eq. (1.8) and the signal placement given by Eq. (1.7). The following sections reveal the practical results of those concepts and what impact they may have on the recovered signal.

2.2.2 Subsampling SNR

As we found out in Chap. 1, the trade-off for this reduction in sample rate shows up as loss in the SNR. Why does the sample rate affect the SNR? Remember that subsampling is actually taking advantage of the phenomenon of aliasing. Any noise in our communication system will also alias into our band of interest, meaning that noise normally out of band is now aliased to occupy the same band as our desired signal. In a conducted environment such as a laboratory there should be negligible out-of-band noise so we will ignore the aliased noise and consider only the noise introduced in the A/D process. Typical A/D noise sources include quantization noise, thermal noise, and sample clock jitter.

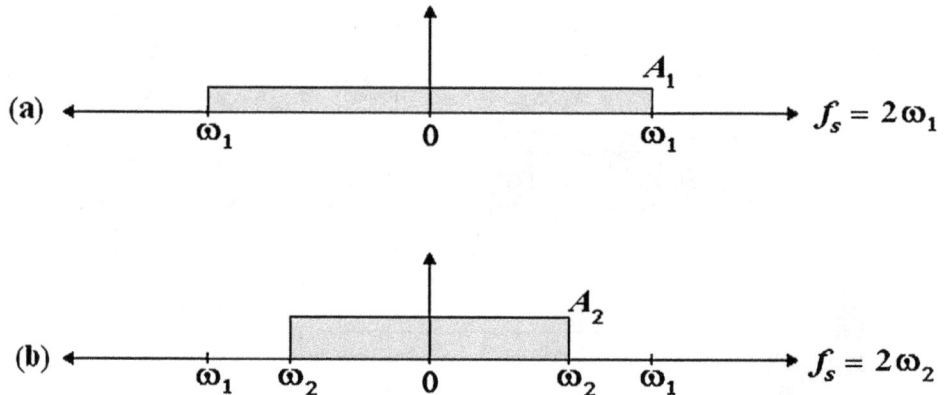

Figure 2.7 Quantization noise power spectral density.

Assuming that all the noise sources are constant, what we really want to know is the change in noise floor that occurs when we lower the sample rate. Oppenheim and Schafer [2] have shown that the quantization noise can be assumed to be a wide-sense stationary white noise process. Such a process has the power spectral density shown in Fig. 2.7a.

Figure 2.7b shows another power spectral density of the same white noise process but with a lower sample rate. In Ref. [2] the authors show that the *total* noise power does not depend on the sample rate. Therefore, the area under the curves in Fig. 2.7a and b is a constant. For the total noise power to remain constant over the smaller range of frequencies in Fig. 2.7b, the amplitude A_2 must be greater than A_1. Conversely, increasing the sample rate reduces the amplitude of the noise to keep the same total noise power over the increased bandwidth and hence one of the benefits of oversampling. For our subsampling receiver, this increased noise amplitude has the effect of raising the noise floor of the A/D. The noise floor in an A/D can be calculated from Ref. [3] as

$$\text{NF(dB)} = 6.02 \times B + 1.8 + 10 \log\left(\frac{f_S}{2}\right) \tag{2.1}$$

where B is the number of bits in the A/D converter and f_S is the sample rate. From Eq. (2.1) the change in NF seen by lowering the sample rate from f_1 to f_2 is given by

$$\Delta\text{NF(dB)} = 10 \log\left(\frac{f_1}{2}\right) - 10 \log\left(\frac{f_2}{2}\right) = 10 \log\left(\frac{f_1}{f_2}\right) \tag{2.2}$$

In fact we can easily see this change in the noise floor by examining Figs. 2.8 and 2.9. Figure 2.8 is the spectrum of a real 16-QAM signal with carrier frequency, $f_C = 99.003$ MHz and a bandwidth of 6 kHz, sampled at 200 Msps, which

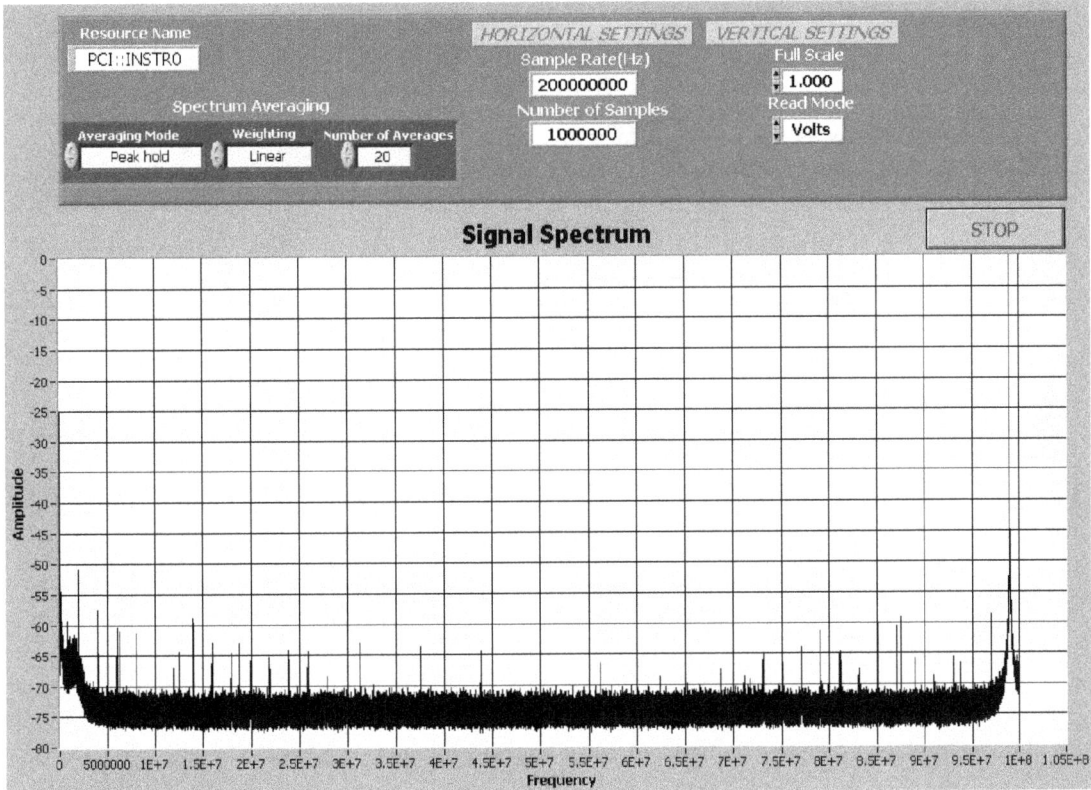

Figure 2.8   16-QAM signal sampled at 200 MHz ($f_C = 99.003$ MHz).

is just above the traditional Nyquist rate. The noise floor in this figure looks to be approximately –70 dB. Figure 2.9 is the spectrum of the same 16-QAM signal subsampled at a rate of 100,000 sps. Clearly the noise floor is now higher than before and from the figure, we can approximate the new noise floor to be around –36 dB. This is an increase of 34 dB in the noise floor level estimating from the figures. Using Eq. (2.2), the increase in noise floor can be calculated as

$$\Delta \text{NF} = 10 \log \left(\frac{200,000,000}{100,000} \right) = 33\,\text{dB} \qquad (2.3)$$

The rise in noise floor just described is the minimum loss in SNR that will be experienced by a subsampling receiver. It is the minimum because here we have excluded any other noise sources such as aliased out-of-band noise or spurs in the A/D conversion process. The effect of the loss of SNR (or an increased noise floor) is an increased probability of error in our demodulated signal. The relationship between SNR per symbol and the probability of a symbol error for

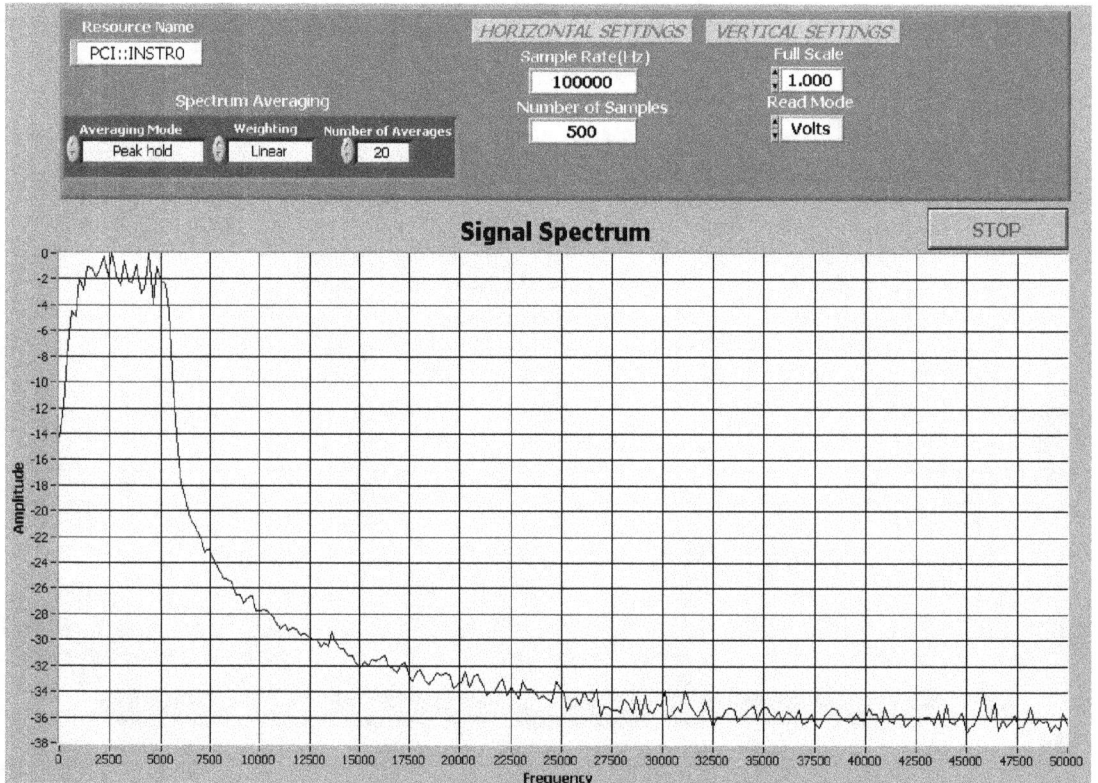

Figure 2.9   16-QAM signal sampled using subsampling at 100 KHz (f_C = 99.003 MHz).

an M-QAM signal is developed in Ref. [4]. For a 16-QAM signal, the symbol error probability is given by

$$P_M = 1 - \left[1 - \frac{3}{2} Q \left(\sqrt{\frac{3}{15} \frac{E_{AV}}{N_0}} \right) \right]^2 \tag{2.4}$$

where E_{AV}/N_0 is defined as the average SNR per symbol.

The $Q(\ )$ function, remember, is known as the complementary error function and has the property of monotonically increasing with its argument. Indirectly, this means that as our signal's SNR per symbol (or E_{AV}/N_0) increases, $1 - Q(\text{SNR})$ decreases and drags down the probability of a symbol error. Now the big question is how do we know when we have undersampled our signal too far and completely destroyed our SNR? One way to know would be to just monitor your bit error rate by transmitting a known sequence of bits and simply adding up the bit errors at the demodulator. A better way would be to make sure that the chosen modulation and symbol rate can accommodate the reduced SNR. In Ref. [5] there is a wealth of information on the subject of how E_{AV}/N_0 relates to the overall SNR and why E_{AV}/N_0 (or really

E_b/N_0) is a good measure of performance for a digital communication system. This relationship is given by

$$\text{SNR} = \frac{E_{AV}}{N_0} \times \frac{R_S}{W} = \frac{E_b}{N_0} \times \frac{R_b}{W} \tag{2.5}$$

where W is the signal bandwidth, R_S is the symbol rate, R_b is the bit rate, E_{AV} is the energy in a symbol, E_b is the energy contained in 1 bit, and N_0 is the noise power spectral density. Now we can work Eq. (2.4) backward to get the required E_{AV}/N_0 for a specified symbol error probability and then solve Eq. (2.5) to give us the necessary SNR for that error rate. As long as we can maintain that SNR, we can transmit data at a given rate and keep our probability of a bit error below our required threshold. Let us take a look at the following example, which illustrates the relationship between P_M and SNR.

Example 2.1: SNR versus Probability of Error   For this example, we will use the signal shown in Figs. 2.7 and 2.8. That waveform was a 16-QAM signal generated by an arbitrary waveform generator with the following parameters: Symbol rate $R_S = 4800$ symbols per second, excess bandwidth factor $\alpha = 0.2$. We want the system to have a *maximum* symbol error rate of 10^{-3}. The first step will be to rearrange Eq. (2.4) to solve for the $Q(\,)$ function in terms of the symbol error probability.

$$Q\left(\sqrt{\frac{3}{15}\frac{E_{AV}}{N_0}}\right) = \frac{2}{3}\left(1 - \sqrt{1 - P_M}\right) \tag{a}$$

Now we substitute for P_M

$$Q\left(\sqrt{\frac{3}{15}\frac{E_{AV}}{N_0}}\right) = \frac{2}{3}\left(1 - \sqrt{1 - P_M}\right) = 0.0003334 \tag{b}$$

Now we use a table for the $Q(\,)$ function such as the one in Ref. [5] to reverse look-up the argument of the complementary error that yields the number shown in Eq. (b). You will notice that the table lists values only to four decimal places, and several arguments of the $Q(\,)$ function yield the same 0.0003 result. We can easily resolve this by choosing the *largest* argument value that will yield 0.0003 based on the assumption that the largest argument value will yield the largest required E_{AV}/N_0 and hence yield the lowest probability of error. So we read a value of 3.48 from the table and now solve the following equation for E_{AV}/N_0.

$$\frac{E_{AV}}{N_0} = (\arg(Q))^2 \times \frac{15}{3} = 60.552 \tag{c}$$

Finally we have to normalize the given SNR per symbol to the symbol rate divided by bandwidth as in Eq. (2.5). To use Eq. (2.5) though, we must first know the bandwidth W. Again we can look at Ref. [5], which gives us the useful formula:

$$W = \frac{1}{2}(1 + \alpha)R_S \tag{d}$$

Now we use $\alpha = .2$, the excess bandwidth, from Eq. (c) and insert Eq. (d) into Eq. (2.5) for W. The R_S parameter cancels and we end up with

$$\text{SNR}_{\text{req}} = \frac{E_{\text{AV}}}{N_0} \times \frac{2}{1.2} = 100.92$$

or approximately 20 dB.

This exercise is mostly academic in that many digital communication texts contain plots of the required SNR per symbol (or per bit) versus probability of a bit error for some of the more common modulation schemes. Rather than calculating this for each system or modulation it is useful to refer to those plots. From this example, we have determined that this 16-QAM system requires 20 dB of signal to noise in order to stay at or below one symbol error in 1000.

At first glance, this section presents two converse views of the relationship between SNR and the sample rate. On the one hand, we have Ref. [2] saying that the total noise power does *not* depend on the sample rate. On the other hand, a 16-QAM signal was presented with a rise in the noise floor and the authors in Ref. [1] tell us that the SNR is degraded in an undersampled signal. But if the total noise power remains constant, *how* can the SNR be affected when it is simply the ratio of signal power to noise power? Figure 2.10a shows a signal with one-sided bandwidth B and with corresponding A/D quantization noise levels A_1 and A_2 at respective sample rates $2\omega_1$ and $2\omega_2$. In order to maximize the SNR in the recovered signal, most receivers would follow the A/D conversion with a lowpass filter to remove any excess noise. Figure 2.10b shows an ideal version of such a lowpass filter with cutoff B. After applying the ideal lowpass filter the resultant signal and quantization noise are shown in Fig. 2.10c. With sample rate $2\omega_1$, the total noise power is given by $A_1 \times 2B$. And similarly, at sample rate $2\omega_2$ the total

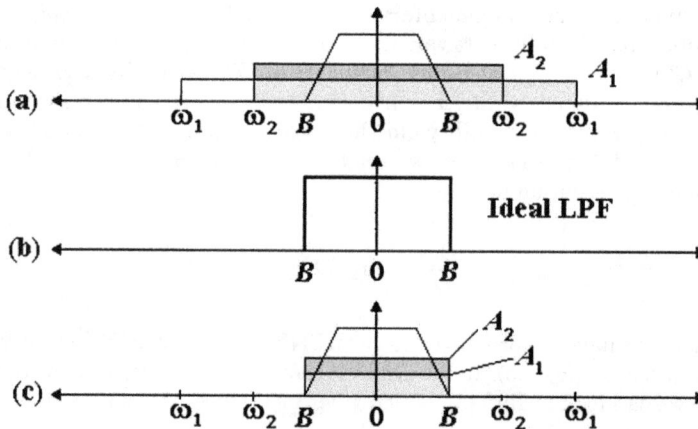

Figure 2.10   Effect of undersampling on total noise power.

noise power is given by $A_2 \times 2B$. Since we know that $A_2 > A_1$, the total noise power after filtering has increased by *lowering* the sample rate from $2\omega_1$ to $2\omega_2$. Assuming that the signal power has not changed, the SNR therefore decreases (since the noise power increases) with the sample rate.

2.2.3 Subsampling signal placement

Another very important aspect of subsampling is the spectral placement of the undersampled translated image. Figure 2.11 shows a zoomed version of the same signal captured in Fig. 2.8. Notice that the signal is centered at $f_C = 99.003$ MHz and the sample rate is 200 MHz. Taking a close look at the subsampled signal in Fig. 2.9 reveals that the signal is now translated to a center frequency of 3000 Hz. Using Eq. (1.7),

$$\text{fix}\left[\frac{f_C}{f_S/2}\right] = \text{fix}\left[\frac{90.003 \text{ MHz}}{100 \text{ kHz}/2}\right] = 1800 \tag{2.6}$$

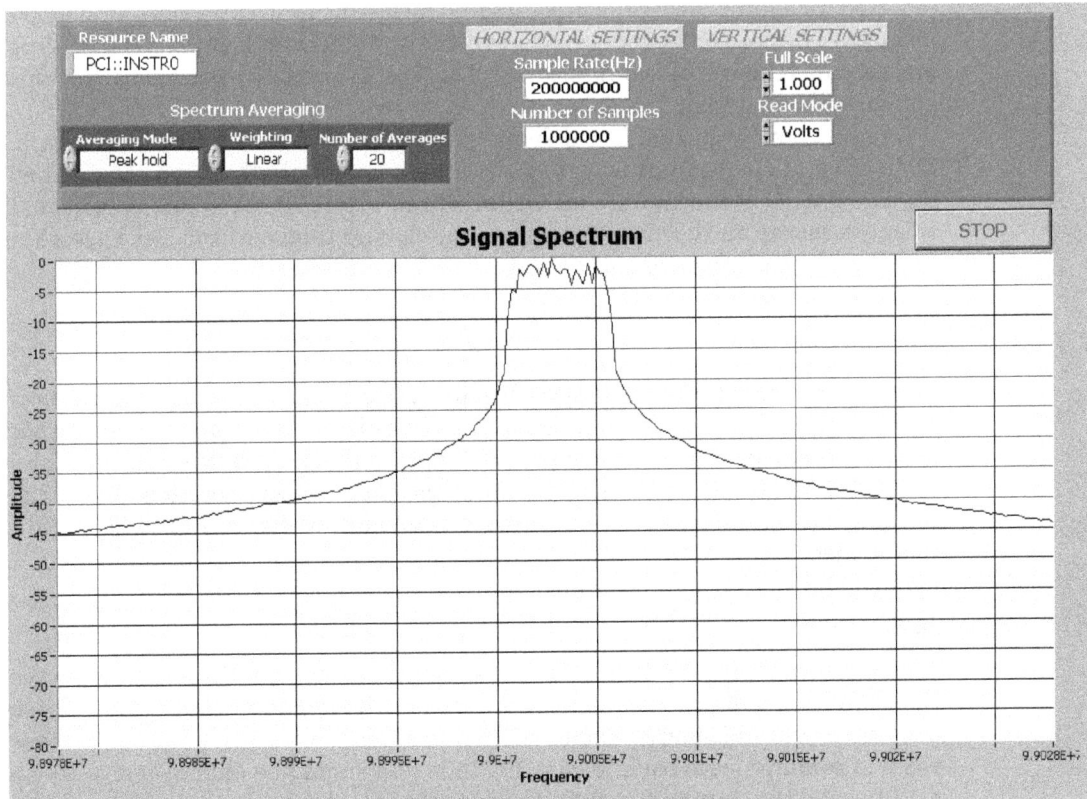

Figure 2.11   16 QAM real signal sampled at 200 MHz ($f_C = 99.003$ MHz).

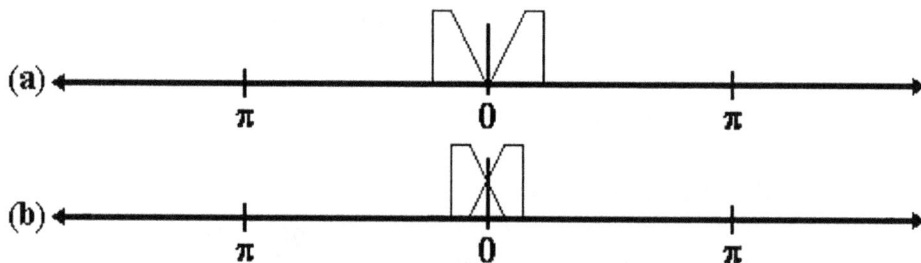

Figure 2.12 Spectral placement of aliased replications.

which is an even number. Then the new IF frequency, f_{IF} is given by Eq. (1.8)

$$f_{IF} = \text{rem}(f_C, f_S) = 3000 \qquad (2.7)$$

If you are paying close attention, you might notice that the carrier frequency was chosen as 99.003 MHz and 0.003 MHz is really what gave us the 3000 Hz offset in f_{IF}. In this configuration, we are now going to have to mix the f_{IF}-centered waveform down to dc in software. The mixing is an extra step, so why not just move the carrier frequency over to exactly 99 MHz and let the subsampling process alias the waveform right to dc? There is a very deliberate reason for this and it is illustrated in Fig. 2.12.

As shown in Fig. 2.11a, a real signal will have an even-symmetric spectrum. If we allow the subsampling process to alias our signal all the way to dc, then the two halves of the signal's spectrum will interfere. It is for this reason that we must choose an IF frequency appropriately and then perform any necessary final mixing in software.

2.3 Other Sampling Methods

The two implementations just described in Secs. 2.1 and 2.2 are not the only possible scenarios for getting your signal into LabVIEW. There are so many other methods that this book could not possibly capture them all. Here some other possible methods are mentioned, but without the elaborate explanation afforded to the previous two methods and the focus is on simply discussing the generalities of each.

2.3.1 Digital oscilloscope

This is a nice clean way to quickly get something working. If you already have LabVIEW and a digital storage scope, you are halfway there. Many scopes these days have built in Ethernet and/or GPIB interfaces. If your scope has one of these ways to communicate back to your PC then you should be able to easily set up a VI to read the data buffer from the instrument. The main limitations of this method will be the bandwidth of the oscilloscope and the speed of the GPIB or

Ethernet for transferring the captured data. There are some instrument specifics such as data format and capture record size that you need to know. You may also have to digitally mix the sampled signal down to baseband in your VI and some methods for actually doing this mix are shown in Chap. 6. Many scopes have only a single sample rate option, typically near 100 Msps. This means *large* amounts of data to work with. One caution is that this method can be extremely slow over GPIB or even Ethernet. Again, this is a rudimentary method, but can get you up and running quickly. The downside here is that the analog bandwidth of the oscilloscope is usually limited to 100 or 200 MHz.

2.3.2 RF spectrum analyzer

This is a very nice method that I like to talk about because a lot of the work is done for you by the spectrum analyzer. Unlike the oscilloscope, most spectrum analyzers have excellent RF specifications and large input bandwidths. Of course, the unit will need to have Ethernet or GPIB capabilities so you can capture the data. Since you set the span and resolution, you know the frequency spacing and therefore the sample rate of the data. Also the data is already in the frequency domain, which may make some filtering operations easier to implement in your processing VIs (see Chap. 9 for tips on filtering in the frequency domain). The key point here is that *you* control the span and resolution, so in effect you are choosing the sample rate. Another thing to remember here is that by setting the carrier frequency, the instrument hardware is mixing your signal to some IF frequency for you. There may be a lot of latency here because of data delay over the GPIB bus and any processing time that the instrument itself may require. But, despite the speed, this is a very easy way to start capturing data and doing some processing in LabVIEW.

2.3.3 Analog sampling card

National Instruments (among many other manufacturers) makes several analog sampling devices. Some are designed to plug into a PCI slot in your PC, others are external boxes with USB interfaces. The advantage of most of these products is the ease with which LabVIEW can communicate with them and the abundance of ready-to-run examples both built into LabVIEW and on the NI website. However, none of the NI cards have analog bandwidths above 150 MHz, which severely limits the range of signals you can analyze without some additional RF hardware to translate the signal to a range that can be captured by the A/D converter.

As shown in Fig. 2.13, LabVIEW has an entire palette dedicated to interfacing with analog input devices. Depending on the manufacturer, these *may* be able to control your particular device—unless the manufacturer has developed his own LabVIEW drivers. These particular functions are fairly easy to implement. There is a configuration to set up the sample parameters such as the number of samples, channel number (for multichannel devices), and type of coupling, and to allocate a buffer for the data. The start-and-read

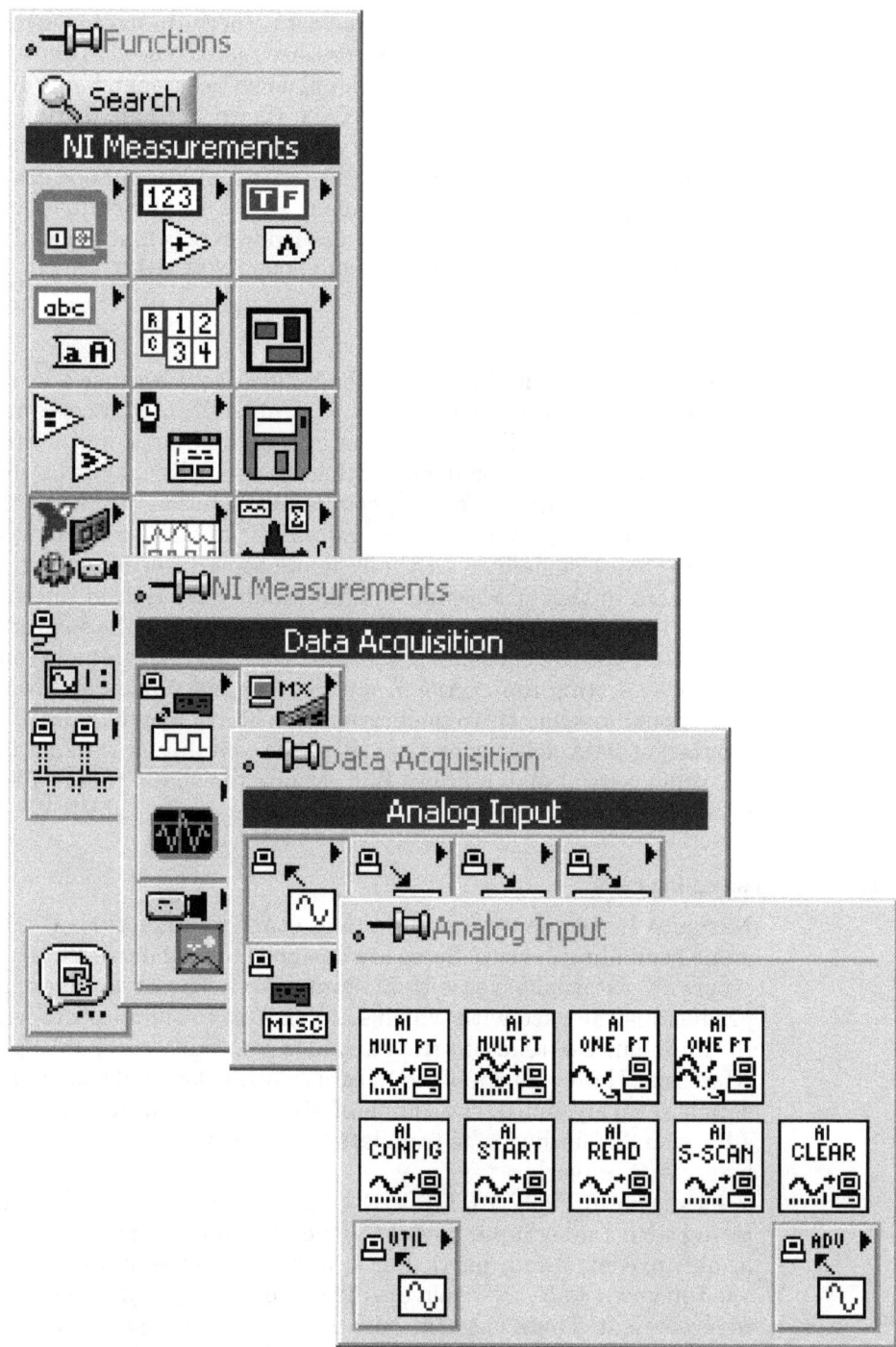

Figure 2.13 LabVIEW analog input palette.

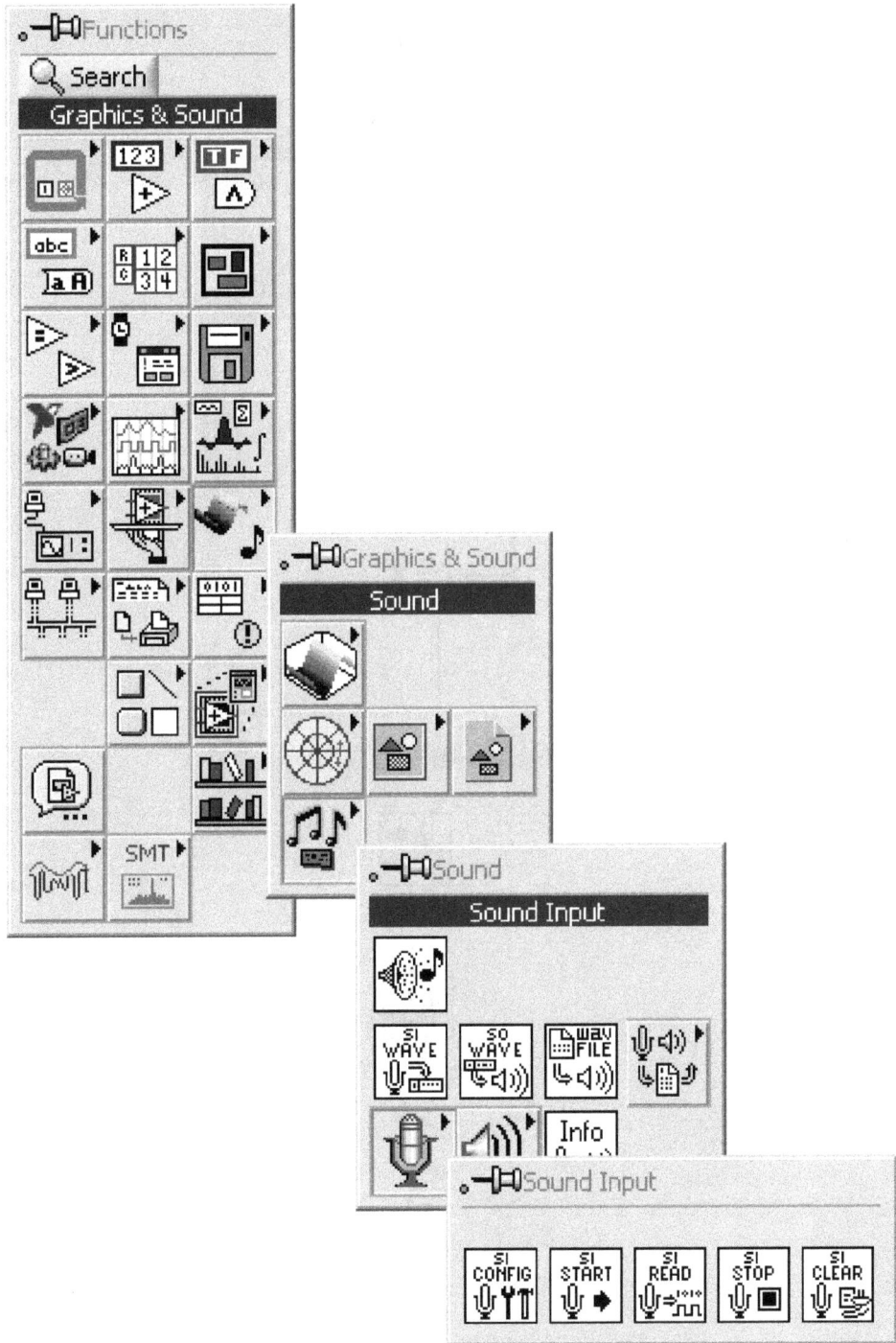

Figure 2.14 Sound input functions palette.

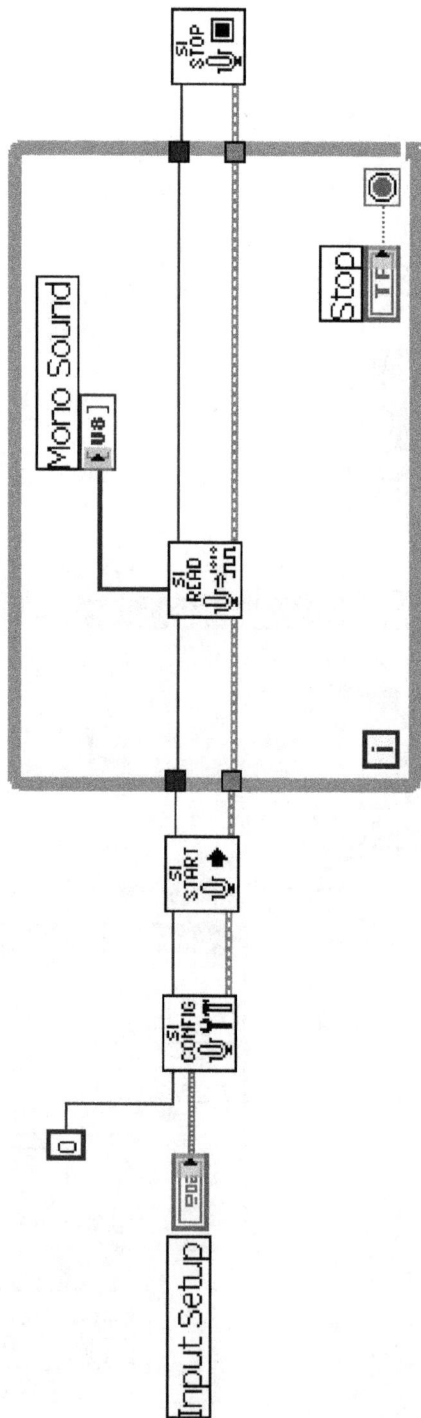

Figure 2.15 SoundCardCapture.vi block diagram.

functions pretty much do what they suggest—start an acquisition and read from the buffer. The actual use of these functions will depend a lot on your particular hardware. The best place for information here will be the user manual and hopefully plenty of LabVIEW examples to get started. Again, please refer to Appendix B for a list of hardware manufacturers and a general description of their offerings.

2.3.4 Soundcard

LabVIEW also has the capability to record data from the audio soundcard installed in your PC (Windows only). There are some limitations to this method, but since almost all PCs now have soundcards, this is a realizable method for sampling real-world signals with LabVIEW. Keep in mind there may be hardware limitations here since the soundcard will have an upper limit on the passband and allowable sample rate, but LabVIEW allows sample rates up to 44.1 kHz with two channels of input and 8 or 16 bits per sample. Of course, the low bandwidth of the soundcard will not allow you to directly capture most RF signals, but you may be able to experiment with sampling other types of signals with this method.

Figure 2.14 shows the location of the LabVIEW sound input functions. As with all of the sampling devices, there is a configuration, a start, a read, and even a stop function. This input method *may* be acceptable if your signal is already at baseband, such as at the output of your transmitter device's DSP.

The block diagram of a generic soundcard capture VI is shown in Fig. 2.15. Here the input setup control allows the user to select the sample rate, bits per sample, and whether the capture is mono or stereo. By default the capture buffer is 8192 bytes in length, although this can be user defined. Once the soundcard is configured, the loop continually reads from the buffer until the user presses stop.

Summary

This chapter has described several methods for acquiring digitally modulated signals using LabVIEW and various types of hardware. As with everything in engineering, there are trade-offs between the types of hardware, the complexity, cost, and performance of each system. National Instruments PXI products are off-the-shelf, LabVIEW-friendly instruments that are really well suited for digital communications. The downside to these products is price and flexibility. In Sec. 2.2, a stand-alone sample card was used to show how to build a completely digital receiver with a single piece of hardware. The obvious advantage here is price, but at the expense of increased complexity. One thing that further chapters will make clear is that no matter how the signal is acquired, the digital processing in LabVIEW does not change and those chapters will focus on building important processing blocks into sub-VIs for incorporation into diverse types of communication systems.

Now that you are more aware of what types of sampling hardware are out there, which method do you choose? Starting from scratch sometimes can be overwhelming because of all the available choices. Choosing to use LabVIEW to process a digital signal should narrow down your choices quite a bit though. Starting with prices, the PXI system starts around $12,000 (excluding the controller) and can easily run over $20,000 depending on your options and configuration. The PXI-5660 does implement undersampling, so it too will have some degradation in SNR, although not as much as with the subsampling receiver. This is simply because the 5660 first mixes the RF to an intermediate frequency and so generally speaking the undersampling factor n will be a smaller number. The Acqiris digitizer used in this chapter costs around $10,000 and is a multi-faceted product in that it is capable of 2 Gsps with 1 GHz of bandwidth, but we can also use it to undersample a signal—we get less control over the PXI devices. Ultimately your budget and your own preferences will have to decide which device is right for you.

References

1. Vaughan, R. G., N. L. Scott, and D. R. White, "The Theory of Bandpass Sampling," *IEEE Transactions on Signal Processing,* vol. 39, pp. 1973–1984, September 1991.
2. Oppenheim, A. V., R. W. Schafer, and J. R. Buck, *Discrete-Time Signal Processing,* 2d ed., Prentice Hall, Upper Saddle River, NJ, 1998.
3. Brannon, B., *Basics of Designing a Digital Radio Receiver: Analog Devices,* 1999. Available at *www.analog.com.*
4. Proakis, J. G, *Digital Communications,* 4th ed., McGraw-Hill, New York, 2001.
5. Sklar, B., *Digital Communications,* 2d ed., Prentice Hall, Upper Saddle River, NJ, 2001.

2

Building Blocks

Spectral Analysis

Before going any further into the processing of digital communication signals, it is important to stop and explore some of LabVIEW's spectral analysis capabilities. These tools will be important later when we examine the spectrum of our input signals in various stages of the communication system. And of course the tools will also be useful when looking at the frequency response of the filters that we design in Chap. 4. Most versions of LabVIEW include all the spectral processing functions you should ever need. Among the standard functions are the real fast Fourier transform (FFT), complex FFT, power spectrum, and the Hilbert and Hartley transforms as well as the inverses of those functions. There are also many expanded spectral functions that build on the aforementioned standard functions. The following sections will develop some of the common standard spectral functions into useful virtual instruments (VIs). In the case where you are required to use your own FFT algorithm or your LabVIEW package does not include the spectral capabilities that you require (such as discrete cosine transforms or wavelet functions), LabVIEW allows for the incorporation of your own compiled C code through a code interface node. This topic is worth mentioning here because of the wealth of C libraries available on the Internet to perform some of these complex spectral operations.

3.1 Low-Level Frequency Domain Functions

Let us start by looking at the basic frequency domain functions shown in Fig. 3.1. Of these functions, probably the most used for digital communications are the complex Fourier transform $F(x)$ and the real Fourier transform $F(x)$. In digital communications, we deal with both real and complex signal types and therefore we will use both the LabVIEW Fourier transform functions. As long as the size of the input signal is a power of 2, these functions implement a split radix FFT, otherwise they perform the discrete Fourier transform. Since the two operations yield identical results, we will generally try to force the more efficient FFT

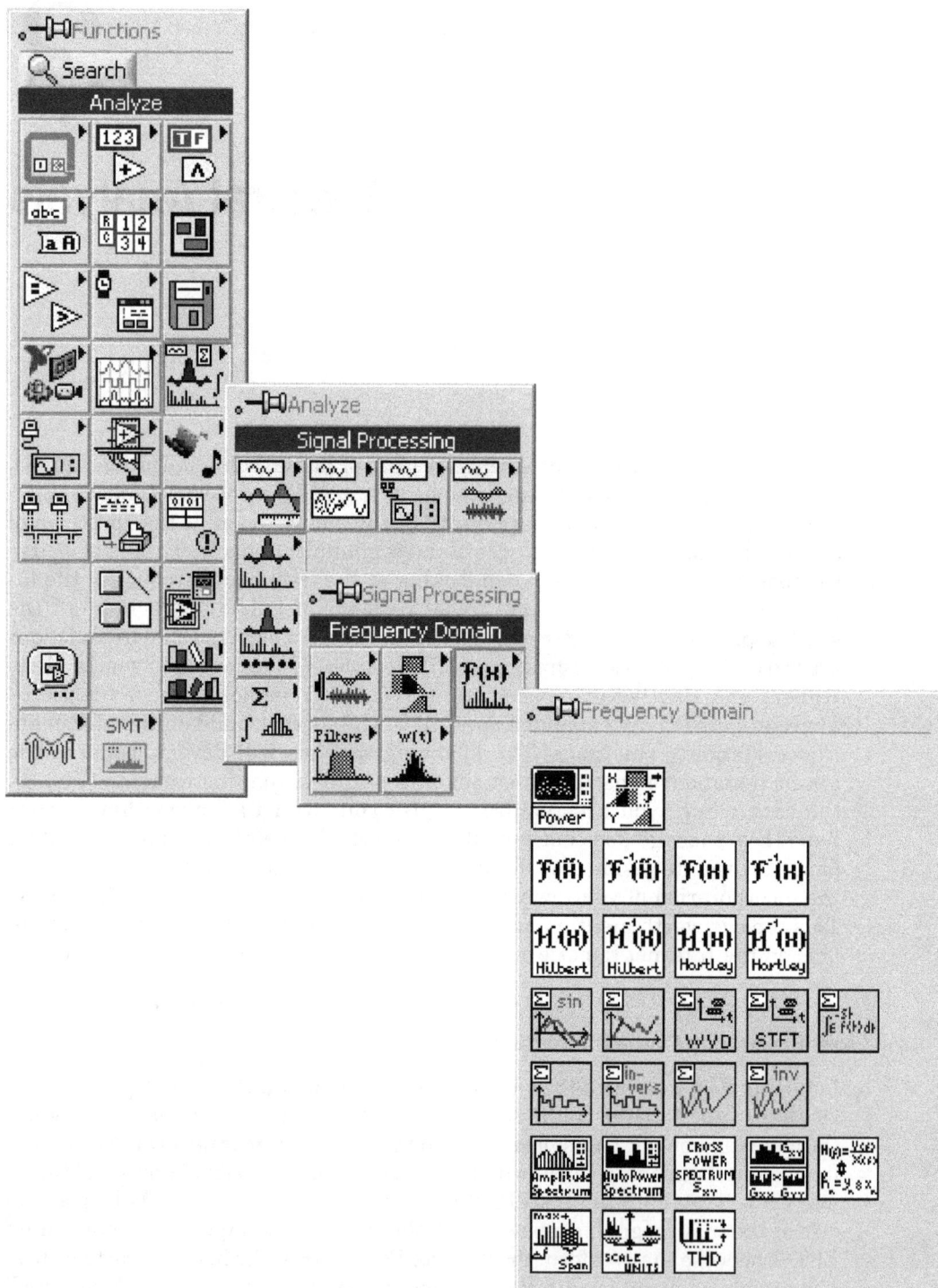

Figure 3.1   Low-level frequency domain functions.

computation by adjusting the input length. Besides operating on complex input, what is the difference between the real and complex versions? Remembering the properties of Fourier transforms, a real signal will have a Fourier transform that is symmetric about 0 Hz in the region from $-f_s/2$ to $f_s/2$. That means that the two halves of the spectrum are mirror images. A complex signal on the other hand will have a Fourier transform with completely independent frequency content in each half band. That being said, the LabVIEW real Fourier transform outputs only the positive half of the signal's spectrum and the complex Fourier transform outputs both halves. The two halves of the spectrum in the complex case are output in the frequency range 0 to 2π, and therefore require some slight shifting in order to view the signal in the range $-\pi$ to π. This sort of operation will be shown in more detail in the next section.

3.1.1   Simple FFT

The simple FFT routine shown at the top of Fig. 3.2 tells us a couple of interesting points about the use of the LabVIEW complex FFT. First of all, the output from the complex FFT starts with the dc component and builds up from there to the half sample rate and then continues onward to the full sample rate component.

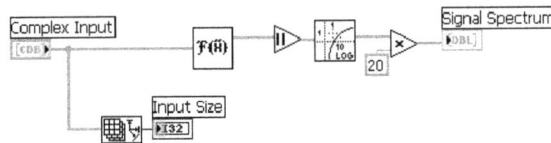

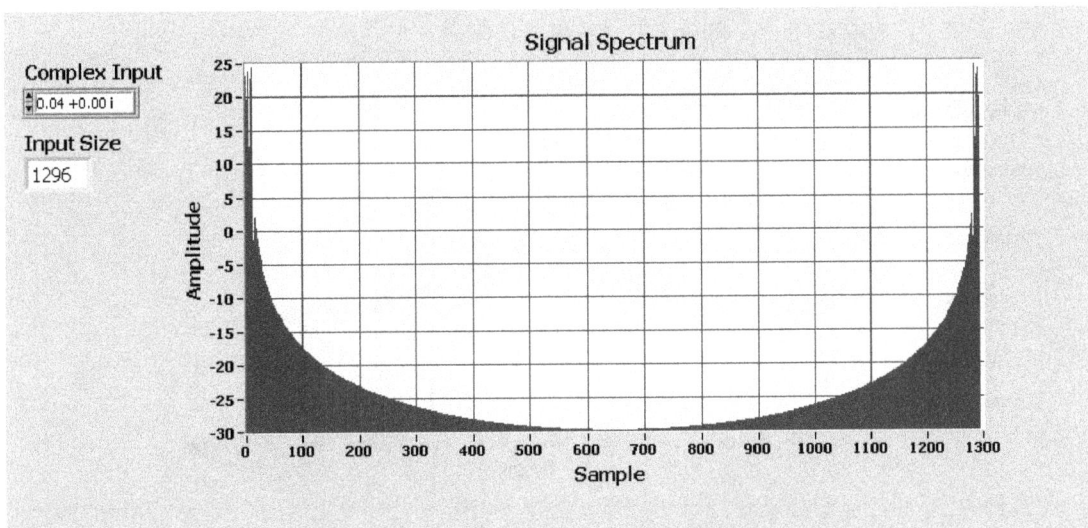

Figure 3.2   Simple FFT computation VI.

This is typical of many FFT outputs and is easily adjusted by splitting the output array and reconcatenating the two pieces as shown at the top right of Fig. 3.3. The other notable feature (or lack thereof) in this example is that the input size is *not* a power of 2. In this case, the complex FFT routine is actually implementing a complex discrete Fourier transform (DFT). If we want to ensure that an FFT is always performed, we will have to pad the data size up to the next power of 2 as shown in the improved VI in Fig. 3.3. Does it really matter whether the function performs an FFT or a DFT? From the discussion in [1], we can estimate that the computation of the DFT generally requires N^2 complex multiplications, where N is the number of points in the DFT. Conversely, the FFT requires only $(N/2) \log_2 N$ complex multiplications. As an example, suppose that the size of the input signal is 200,000 samples. If we want to examine the spectrum of the entire signal by using the DFT, it requires 40,000,000,000 complex multiplications. That is quite a lot of operations for any computer to perform. By using the power of the FFT, we can cut the number of complex multiplications down to about 1,760,000. There is a difference of more than a factor of 10,000 between the two implementations, and thus we can easily see the advantage of using the FFT whenever possible.

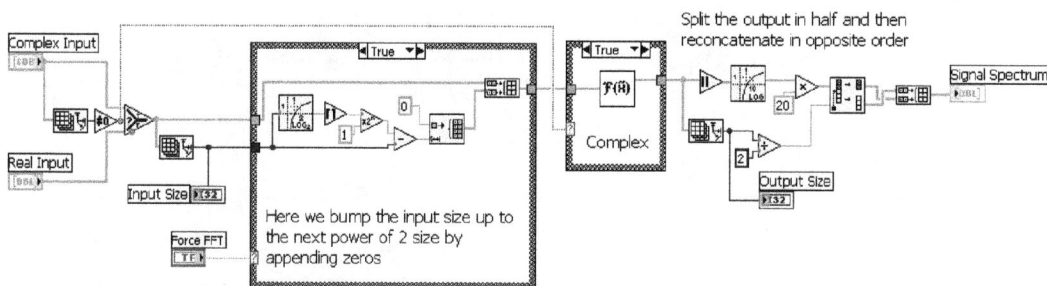

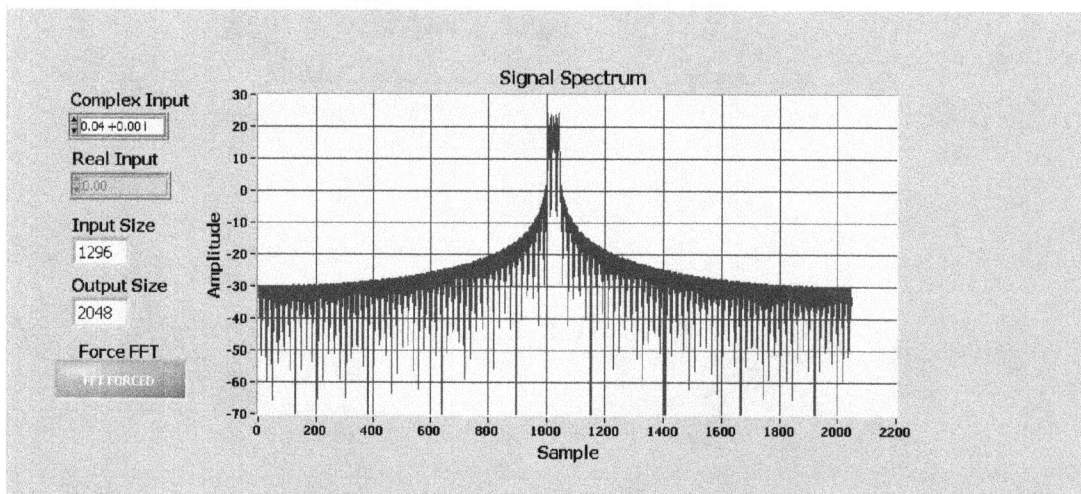

Figure 3.3 AdvFFT.vi.

3.1.2 Improved FFT

Figure 3.3 shows the block diagram and front panel for an improved FFT VI called AdvFFT.vi. As mentioned previously, this function now has the capability to force the LabVIEW FFT routine to *always* compute the FFT by appending the appropriate number of zeros to the input in order to bump N up to the next power of 2. Notice too that the spectrum displayed on the front panel is now altered so that it has the property of being centered at the dc component with increasing negative frequencies to the left and increasing positive frequencies to the right. In addition to these features, AdvFFT.vi also has the ability to accept real or complex input and will call the corresponding FFT function. Now we have a pseudopolymorphic input VI to handle the real or complex FFT and plot the resulting shifted spectrum for us. We will call upon this particular VI regularly to view the signal at various points in the communication system as well as to examine the output of the filters in Chap. 4.

At this point, we will really need one more piece of the puzzle to finalize the use of AdvFFT. That piece is the generation of the values for the frequency axis. As is, we have no knowledge of the relationship between the signal components plotted in Fig. 3.3 and their corresponding frequency values. Recall that the DFT is computed by [1]

$$X(k) = \frac{1}{N}\sum_{n=0}^{N-1} x(n)e^{-j\frac{2\pi nk}{N}} \qquad k = 0,\dots,N-1 \qquad (3.1)$$

Using Eq. (3.1), we can see that the DFT is calculating the frequency content of the input signal $x(n)$ at N equally spaced frequencies determined by $e^{-j(2\pi k/N)}$ for $k = 0, \dots, N-1$. That means then that there will be N discrete frequencies (typically known as DFT bin frequencies) where the frequency content of $x(n)$ will be evaluated. Keep in mind that if we were to compute the continuous Fourier transform of a signal, we would evaluate the frequency content of the signal at an infinite number of frequencies. And so for this reason, many sources refer to the DFT as a sampling of the continuous Fourier transform. Because we only *look* at the frequency content of $x(n)$ at those N discrete frequency values, any spectral content of $x(n)$ *not* at those discrete values will not be properly accounted for. Is there anything we can do about this? One way to alleviate part of this problem is to put those N discrete frequencies closer together by increasing the value of N. This is easily done by zero padding the input signal as was done in the AdvFFT VI to force the FFT function. So here we get two benefits of zero padding: (1) the speed improvement of the FFT and (2) the increased spectral resolution by sampling the continuous Fourier transform at a higher rate. As always, there will be a trade-off between the spectral resolution that we desire and the time it takes to compute the FFT. And so finally, this brings us to the question, how exactly do we relate the $e^{-j(2\pi k/N)}$ spacing of the DFT to frequencies that make sense to us? Here we absolutely *must* know the sample rate of the signal $x(n)$. Assuming $x(n)$ is sampled at f_s, we can determine

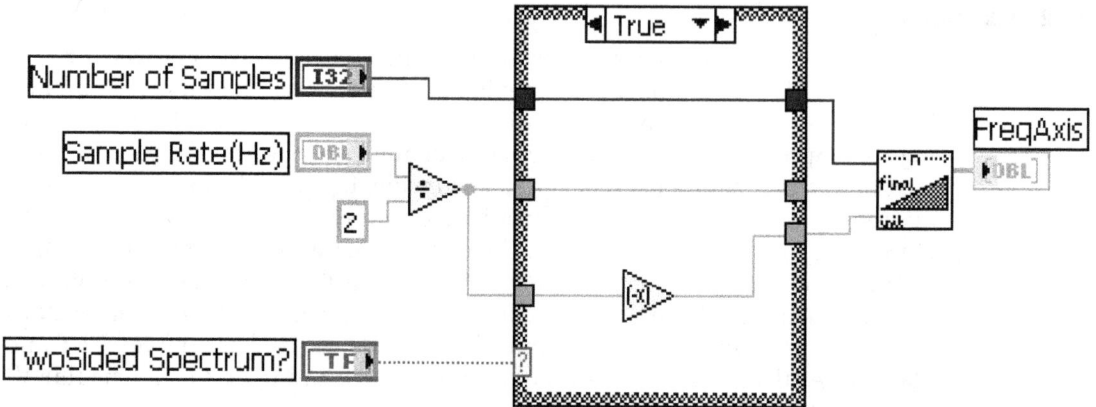

Figure 3.4 FreqAxis.vi generates the frequency axis based on Eq. (3.2).

the frequency spacing between successive samples of the DFT to be

$$\Delta f = \frac{f_s}{N} \text{Hz} \qquad (3.2)$$

Using Eq. (3.2), we can easily generate the frequency axis for our plotted spectrum in AdvFFT. The VI shown in Fig. 3.4 is called FreqAxis.vi and its only function is to generate the frequency axis values based on the spacing between samples given in Eq. (3.2). The inputs are N (the number of samples), f_s (the sample rate), and a Boolean to determine whether to generate a single-sided or two-sided frequency axis.

With the addition of the FreqAxis VI, the final block diagram for AdvFFT.vi is shown in Fig. 3.5. The signal spectrum plot is changed from a waveform graph to an XY graph in order to accommodate the two-axis inputs. Now if we edit the connector pane to tie complex input, real input, sample rate, and force FFT to the input terminals and signal spectrum to one of the output terminals we can use AdvFFT in other applications as a sub-VI.

3.2 Analyzing the DFT Results

Just as important as being able to use the DFT is being able to understand the results that you get. Earlier, I mentioned that any spectral components of the input signal $x(n)$, *not* precisely at one of the N DFT frequency bin centers, will not be properly accounted for in the DFT computation. The interesting things about any spectral component away from the bin center frequency are: (1) that component's amplitude will not be correctly accounted for because of the shape of the sampling window at that DFT bin, known as scalloping loss and (2) there will be energy from that component in *all* other DFT bins, known as spectral leakage [2]. Let's take these two astounding facts one at a time starting with the latter.

Split the output in half and then reconcatenate in opposite order

Here we bump the input size up to the next power of 2 size by appending zeros

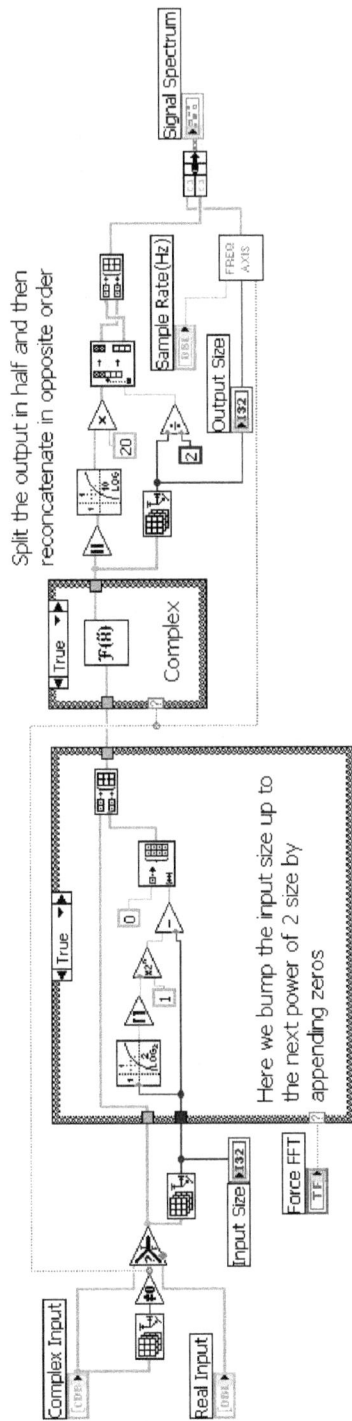

Figure 3.5 Final form for advFFT.vi.

3.2.1   Spectral leakage

Spectral leakage is a term used to describe the phenomenon of energy from one DFT bin leaking into another DFT bin (or more generally into many other DFT bins). Leakage happens because of the lack of orthogonality between some frequency components in our signal and the set of basis vectors in the DFT [2]. The DFT is essentially a computation of the projection of the input signal $x(n)$ onto the orthogonal DFT basis set made up of the sines and cosines at N discrete frequency values evenly spaced from 0 to f_s. It turns out that the projection of any spectral component of $x(n)$ not *exactly* at one of those N discrete frequency values in the DFT basis set will have nonzero projections on *all* frequency values in the basis set [2]. This all means that unless we do something to reduce the spectral content of $x(n)$ at non-DFT bin center frequencies, the results of the *entire* DFT calculation could be grossly inaccurate. And the way we reduce any part of spectral content of a signal is to use a filter. Interestingly, there is a class of filters commonly used for just this purpose known as windowing functions. The following section discusses exactly what is meant by windowing.

3.2.2   Sampling window shape

You might be saying to yourself, "What sampling window?" And the truth is that even *no* sampling window is a sampling window. What is important here is that a rectangular sampling window is implied even when you do not want a window. The implied window comes from the fact that the DFT is a finite sum.

Considering the windowing, we can enhance the DFT equation from Eq. (3.1) to include the window term $w(n)$

$$X(k) = \frac{1}{N}\sum_{n=0}^{N-1} x(n)w(n)e^{-j\frac{2\pi nk}{N}} \qquad k = 0, \ldots, N-1 \qquad (3.3)$$

We know from the properties of Fourier transform, that multiplication in the time domain is circular convolution in the frequency domain. We can thus interpret the time domain window as a filter in the frequency domain. In this case, a window is really nothing more than a filter centered at each DFT frequency bin. This filter provides much needed amplitude reduction for any spectral component of $x(n)$ away from the bin center. Although the rectangular window has a very narrow main lobe, its major problem as a filter is that the side lobes are only 13.5 dB down and do not do a good job of minimizing the impact of the spectral leakage described previously. Since we have no choice but to use a window, we might as well choose a decent window. There has been much research in the design of windows for the DFT (among other uses), and as a result there are many windows to choose from—Hamming, Hann, Blackman-Harris, and more. Typically, the trade-off is in the height of the side lobes versus the width of the main lobe. The window that you choose should be the best window suited to your signal. A window that may work particularly well for speech processing may not be as desirable for digital communications. Some windows have a pedestal while some taper to zero at the edges. All the various windows have subtle differences, with the main classification being side-lobe attenuation and main-lobe width. For a thorough listing of window side-lobe heights and main-lobe widths along with a more detailed explanation, please refer to [3]. Before we go on, let us briefly take a look at Fig. 3.6 to get a visualization of what this windowing filter is doing in the frequency domain.

Figure 3.6 shows the spectra of two different window functions, the rectangular window (dashed line) and the Hann window (solid line). The frequency axis in the figure is shown as multiples of the base DFT analysis frequency f_s/N. As mentioned previously, the multiplication by a time-domain window becomes a circular convolution in the frequency domain. That means that the response in Fig. 3.6 will be centered at each discrete analysis frequency (nf_s/N, for $n = 0, \ldots,$ $N - 1$) in the DFT. We can clearly see that the main lobe of the Hann window spans a whole other DFT bin (one extra bin on each side when looking at the two-sided case), but the side lobes fall off rather quickly compared to the rectangular window. For the DFT computation, this means that any signal energy contained in the next bin higher and the next bin lower gets attenuated slightly and added into the DFT component calculated for a given bin (here is the cause of spectral leakage). The upside to the Hann window (or any nonrectangular

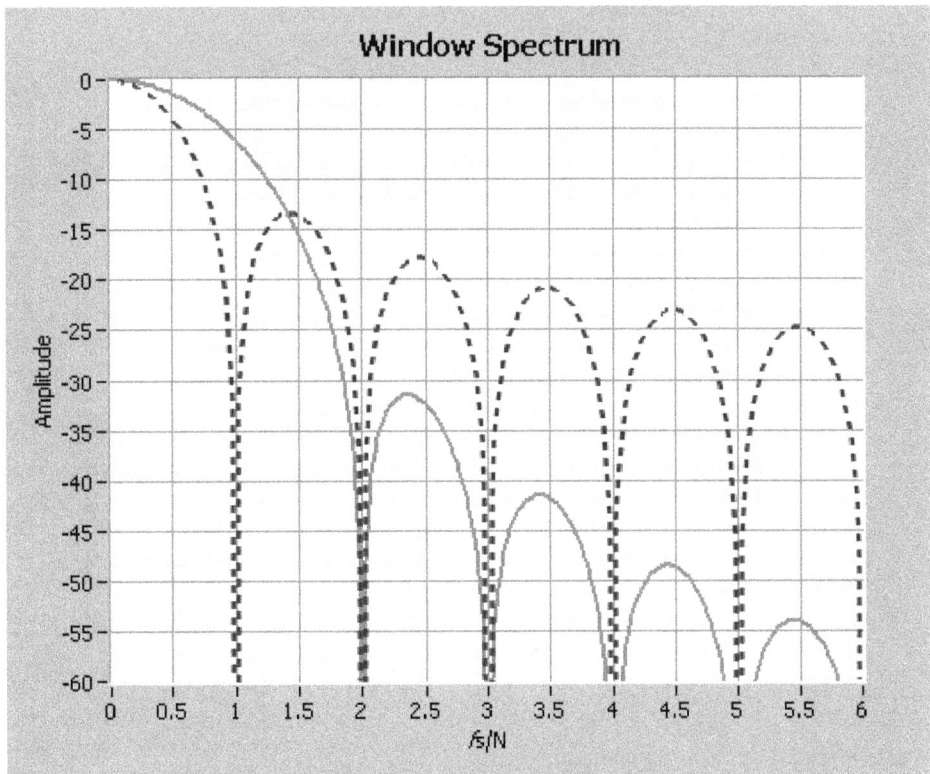

Figure 3.6 Spectra of a rectangular window (dashed) and a Hann window (solid).

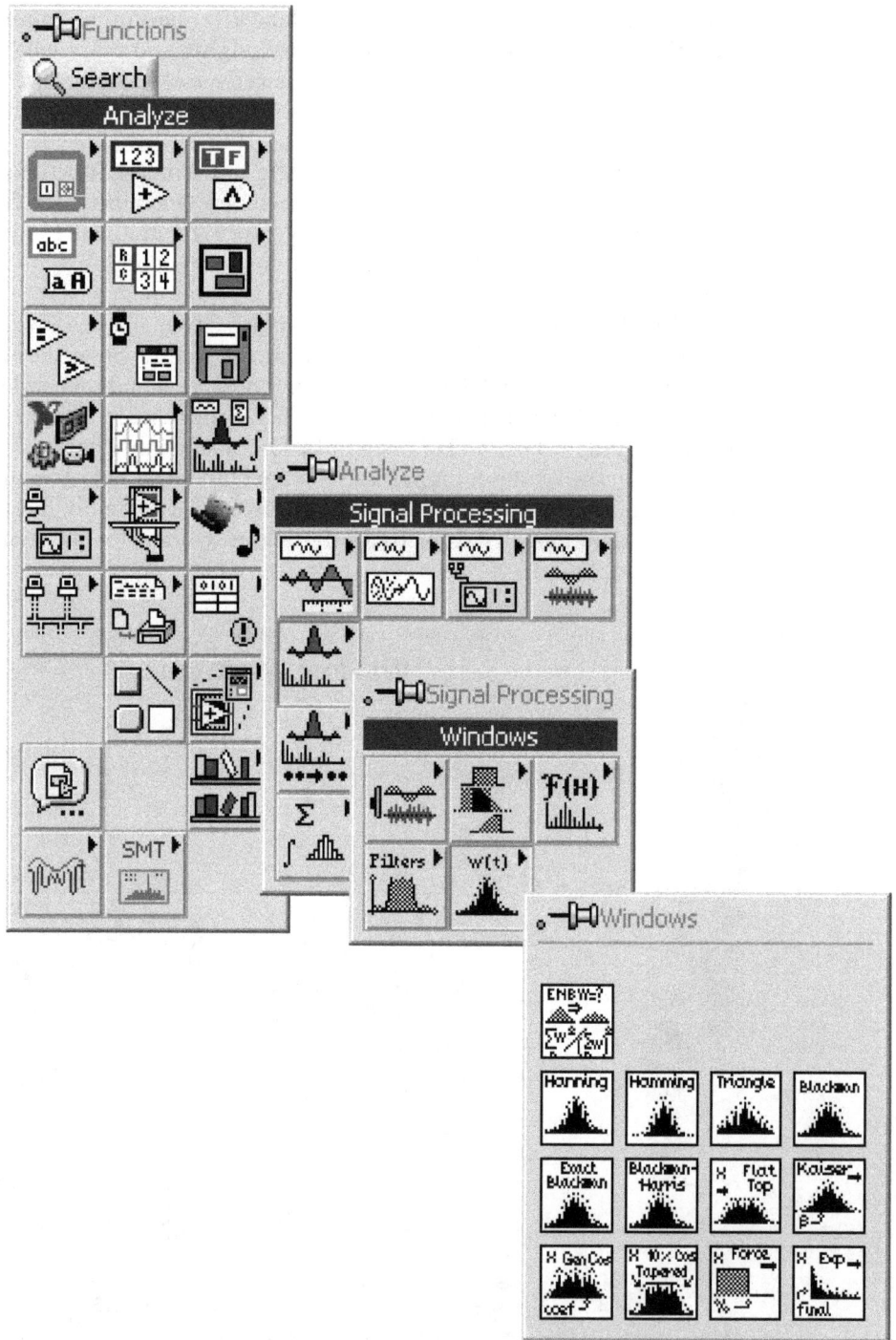

Figure 3.7 LabVIEW windowing functions.

window) is that the energy contributions from bins further than one bin away are attenuated more than those from the standard rectangular window. And finally the concept of scalloping loss is also illustrated in the figure. Remember from the earlier discussion that signal amplitudes away from the bin centers are *shaped* by the sampling window. Using the rectangular window as an example, a frequency component halfway between two bin centers will be attenuated by almost 5 dB. In reality does it really matter if the DFT output is attenuated at some frequencies? Well, this could be important in applications such as speech processing where the amplitudes of certain frequency components need to be measured precisely. The important thing is to keep these degradations in mind when using the DFT.

Now that we understand the DFT results a little better, let us take a look at using windowing in LabVIEW to improve the DFT computation in AdvFFT.vi. Figure 3.7 shows where the LabVIEW window functions are located. LabVIEW has already included some of the most common window functions in signal processing: Hann, Hamming, Triangle, Blackman, Blackman-Harris, and Kaiser. Of course you can always add your own algorithm if you require a specialized window function by using the following examples and substituting your own window coefficients.

If you look at the block diagram of any of the window functions in Fig. 3.7, you will see that they are all essentially the same. Each one of them calls the sub-VIs named General Cosine Window.vi and windowcoefs.vi. All of the previously mentioned windows are formed by a summation of weighted cosines. Because of that, all we really need are the coefficients, which are simply constant arrays chosen by a case selector within windowcoefs.vi. We can make use of this sub-VI ourselves and insert it into a block diagram as shown at the top of Fig. 3.8. By windowing our input before sending it to AdvFFT.vi we can reduce the spectral leakage in the DFT computation *and* also improve the ability of the DFT to resolve two close frequencies. The figure shows the spectrum of the input signal with no window applied (i.e., rectangular window) and with a Blackman-Harris window applied. The dashed line is the rectangular windowed signal and the solid line is the same signal with a Blackman-Harris window applied. There are a few observations that we can immediately make about the effect of the Blackman-Harris window. First of all, the peak magnitude of the windowed spectrum is 5 to 7 dB less than the nonwindowed case. Secondly, the windowed signal (solid line) is noticeably smoother than the nonwindowed signal. In this case, our Blackman-Harris window has side lobes that are down by approximately 61 dB. That side-lobe suppression has cleaned up the signal shown in the figure and provided us with a smoother spectral plot.

So now we know a little bit more about the window functions and we know how to generate them in LabVIEW. The big question is which one to choose. Earlier we talked about the basic trade-off, side-lobe attenuation for main-lobe width, but those may not be the only factors for you to consider. Some windows may have trailing side lobes that decay at 6 dB per octave and some may decay at 18 dB per octave. The important thing to keep in mind here is that whichever window we choose, it should be better than the rectangular window. It is also

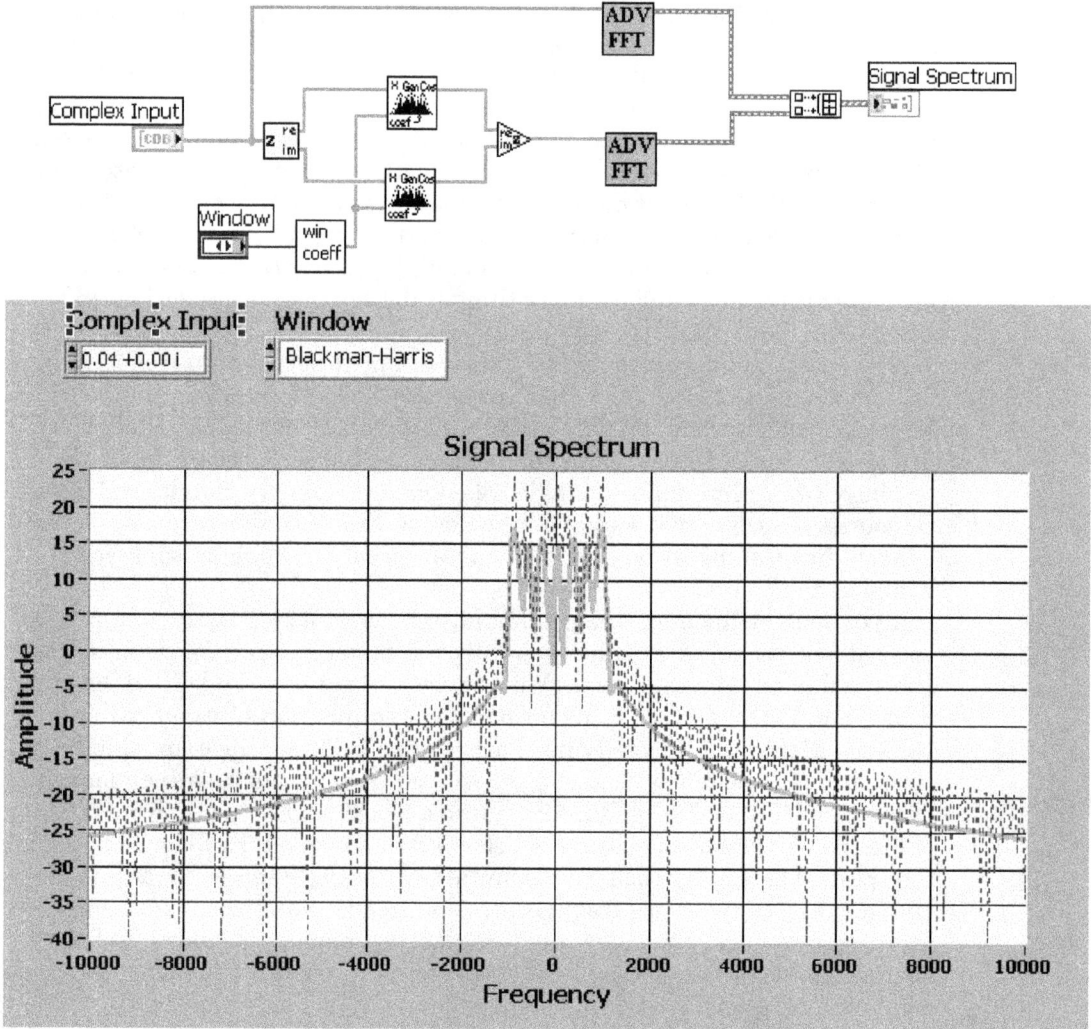

Figure 3.8 Effect of windowing with rectangular window (dashed) and Blackman-Harris window (solid).

advised that the reader consult [2] for a very thorough comparison of the various windows.

3.3 High-Level Spectral Functions

In addition to the low-level signal processing functions described in Sec. 3.1, LabVIEW also includes some useful higher-level VIs, which extend the basic functionality of the low-level spectral tools with features such as windowing, averaging, and density calculations. The location of these functions is shown in Fig. 3.9.

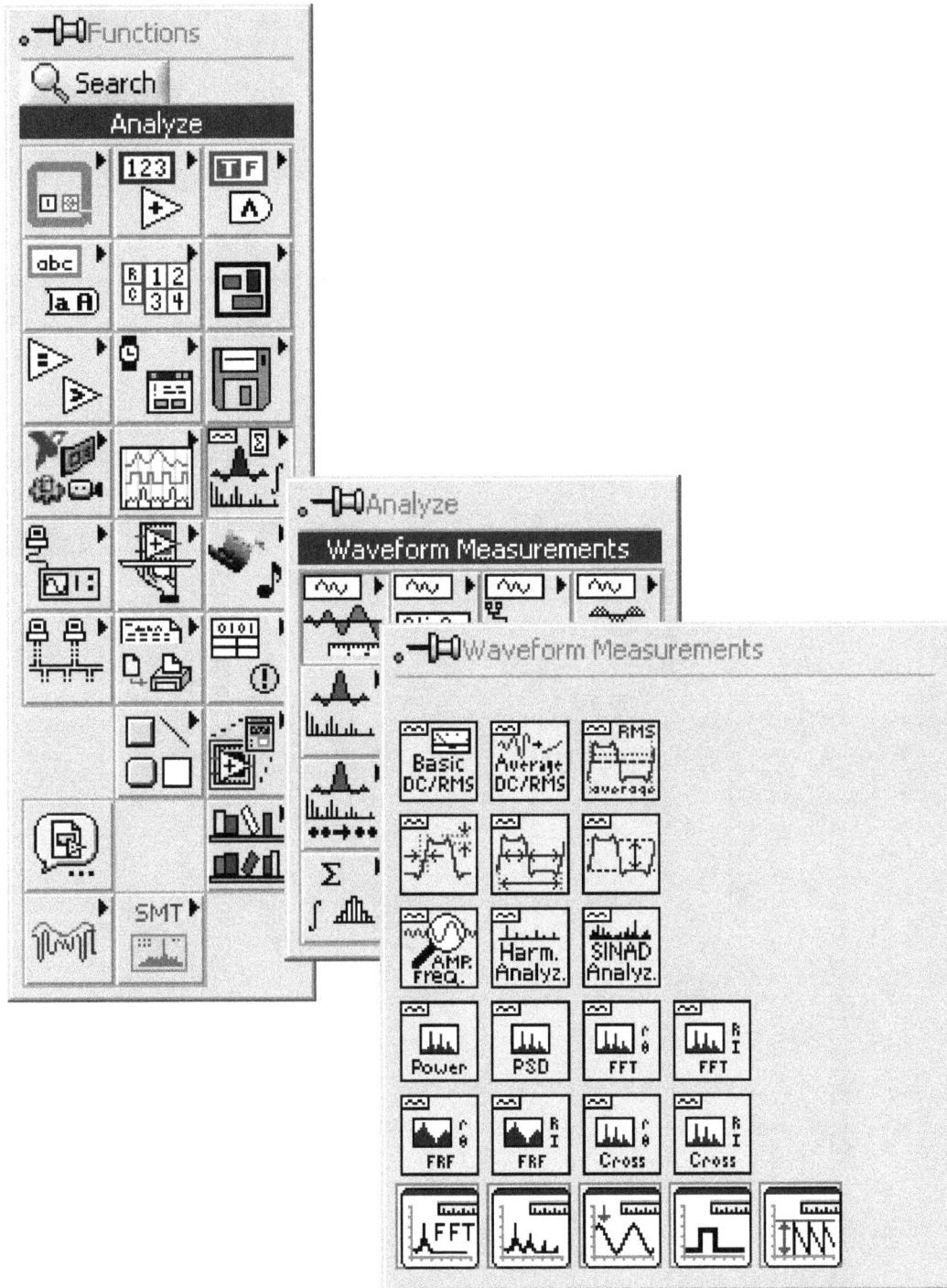

Figure 3.9 High-level waveform measurement tools.

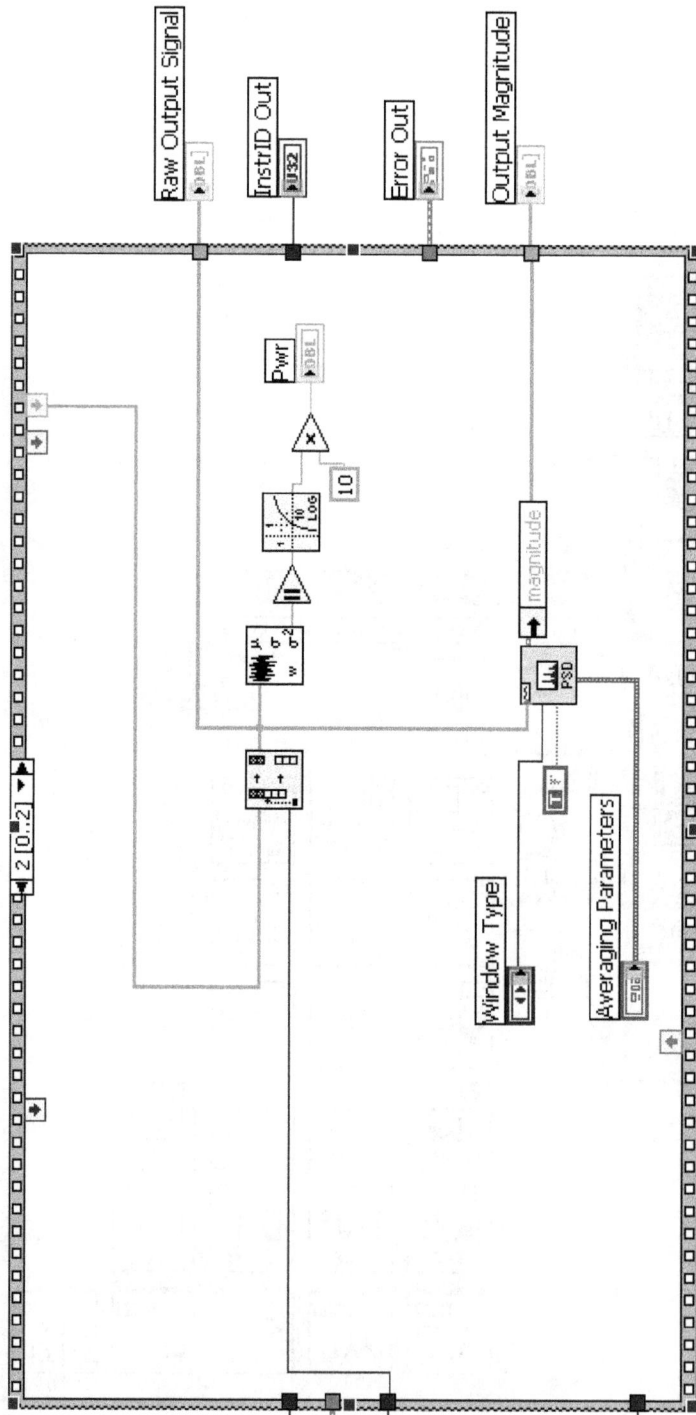

Figure 3.10 SampleLoop.vi shown with power spectral density function.

If you break down the higher-level functions under the waveform measurements subpalette, you will see that at the lowest level they are all built around the FFT or the Power Spectrum VIs shown in Fig. 3.1. You will have to drill down several layers to see those basic functions, but they are the core of each of these high-level functions. The rest of the LabVIEW code handles some formatting and unbundling clusters and user control. So do you want or really need all that extra code? Well that certainly depends. All of these LabVIEW functions are designed to get you up and running as quickly as possible and to that end the NI functions are usually written in a generic fashion to support various uses.

Most of the functions mentioned previously seem written to support quick measurements on signals captured with an A/D card. That means that those signals will be real valued and therefore these functions typically accept only real-valued time-domain inputs. Once we multiply our real-valued signal by a complex mixer, we convert it to a complex signal and many of these functions become unusable from that point forward in our communication system. Chapter 2 showed some examples of real signals captured using an A/D card. The spectrum of those signals was displayed on the front panel using the power spectral density VI as shown below in Fig. 3.10.

Some of the nice features of the waveform measurement functions are the weighted averaging, peak hold, and windowing capabilities. These features are especially useful if you want to simply display or record direct measurements on the input spectrum without any further processing. As with all of the VIs developed by National Instruments, these tools are especially helpful as programming guides for building your own special purpose spectral tools.

3.4   Adding C Routines to LabVIEW

This section will briefly explain ways to expand the intrinsic capabilities of LabVIEW with your own C functions. The reason this topic is included here is because many engineers may have their own tried and true C function for computing the discrete cosine transform or wavelets or perhaps some other type of transform that is specific and proprietary. This by no means implies that only spectral processing C functions should be incorporated into your VI, only that these types of algorithms are typically computationally intense and are commonly optimized in C. There are also other reasons that you may want to include a C function within LabVIEW such as speed, compactness, and even ease of development. Whatever the case may be, there may come a time when you need to include that function in LabVIEW. With that in mind, the following paragraphs will outline the basic steps for incorporating a C routine for computing the FFT.

The first thing you will need to get started is the appropriate dynamic link library (DLL) for the function that you wish to integrate into LabVIEW. For our purposes, I did a Web search for FFT libraries and came upon one which was straightforward and included a DLL with a very simple interface. The one I chose was written by Murphy McCauley and can be downloaded from *www.fullspectrum.com/deeth*.

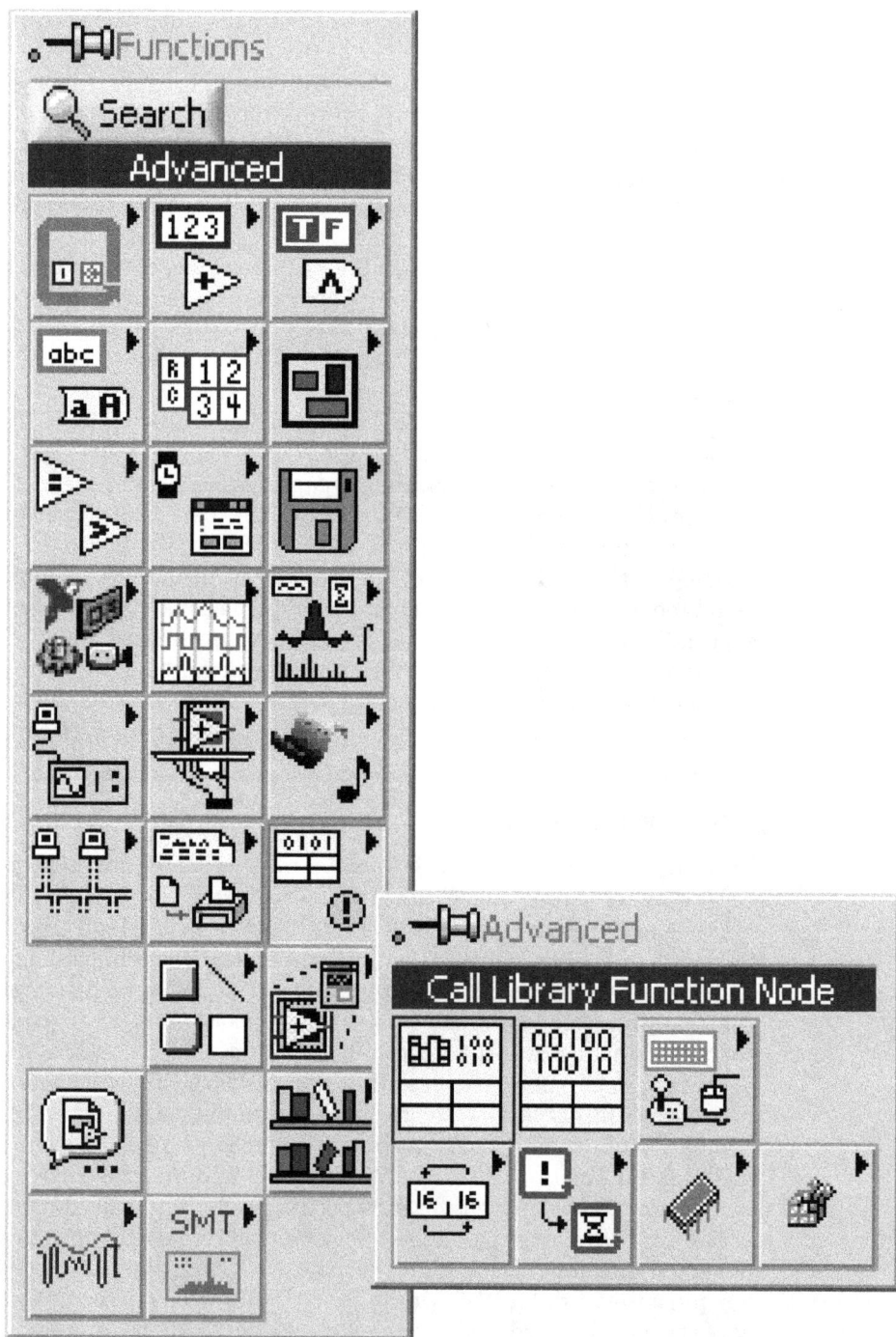

Figure 3.11 Advanced functions palette.

To start incorporating this C function into LabVIEW, the first step is to examine the function prototype in the header file, in this case FFTLib.H. In reading through the provided documentation for this DLL, you may notice that the author provided some wrapper functions called fftDouble and fftFloat to load the DLL and call the necessary functions fft_double( ) and fft_float( ). In our case, LabVIEW will handle loading the library, so we can eliminate any calls to the wrappers and go straight to the real function, fft_double( ). You can weed through the FFTLib.c file and see that the function fftDouble( ) simply passes all of its arguments to fft_double( ). With that in mind, we can determine how to call fft_double( ) by looking at the function prototype excerpt from the following FFTLib.h.

Code Excerpt from FFTLib.h by Murphy McCauley

BOOL fftDouble (

Unsigned	NumSamples,	/* must be a power of 2 */
Int	InverseTransform,	/* 0 = forward FFT, 1 = inverse FFT */
Double	*RealIn,	/* array of input's real samples */
Double	*ImagIn,	/* array of input's imag samples */
Double	*RealOut,	/* array of output's reals */
Double	*ImagOut);	/* array of output's imaginaries */

Now that we know how to call the FFT function, we are ready to work in LabVIEW. The first thing to do is to insert the call library function node into the VI. This node is found in the advanced subpalette as shown in Fig. 3.11.

After inserting the call library function node, right-click on it and choose configure from the drop-down menu as shown in Fig. 3.12. This will pop up a window that allows you to define the function interface in a way that LabVIEW can understand. Most of the required information is straightforward. Obviously you will have to point LabVIEW to the correct DLL and type the name of the function appropriately. The tricky part is making sure that you set up all the input parameters according to the function prototype shown in the code excerpt given previously.

Notice that as you add each parameter using the add-a-parameter-after button, LabVIEW automatically builds the function prototype in the bottom window. A quick look at this prototype should confirm whether you have entered the function information appropriately and then you can start wiring up the inputs and outputs.

A completed FFT DLL VI is shown in Fig. 3.13. The inputs are number of points, transform direction, a pointer to the real part of the input, and a pointer to the imaginary part of the input. LabVIEW handles the form of the passed arguments based on the function prototype and all the user needs to do is to wire the appropriate signals to the terminals. As for the outputs, space is allocated for the real and imaginary arrays by initializing two arrays with 0s.

The ability to incorporate DLLs directly into a VI makes signal processing in LabVIEW extremely versatile. Typically, engineers will have custom routines written in C that are algorithmically intense, but do not necessarily

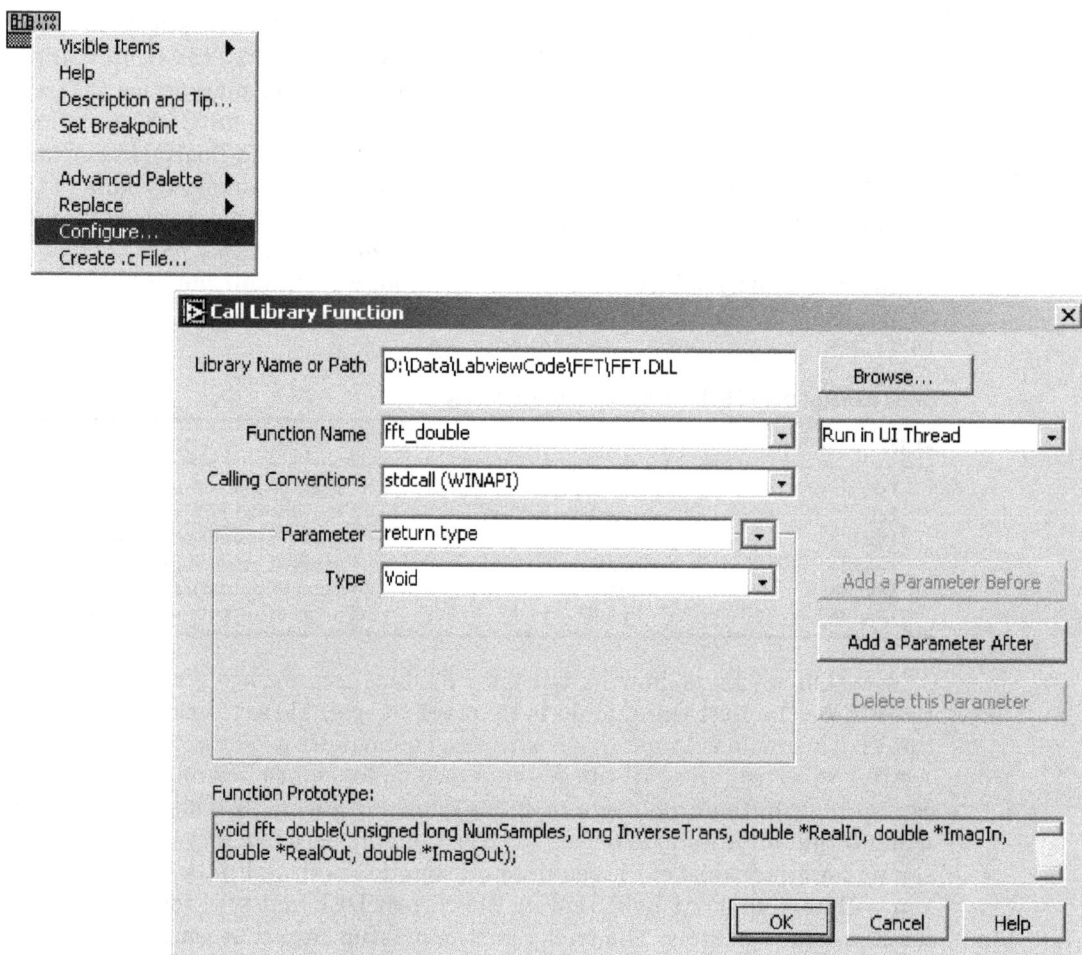

Figure 3.12 Configuring the call library function node.

have a decent user interface. By using LabVIEW for the control and graphics capabilities, many powerful signal processing tools can be built around custom DLLs.

3.5   Spectral Measurements Toolset

At this point it is appropriate to discuss one of the optional National Instruments Spectral Measurements Toolset. This suite of tools is shipped as part of the PXI-5660 RF spectrum analyzer package discussed in Chap. 2. The toolset contains functions for measuring in-band power, occupied bandwidth, adjacent channel power as well as finding spectral peaks and signal spectrum averaging. There are also some very handy example VIs, which will capture and plot the spectrum of your signal and allow you to set the carrier frequency, span, and averaging type.

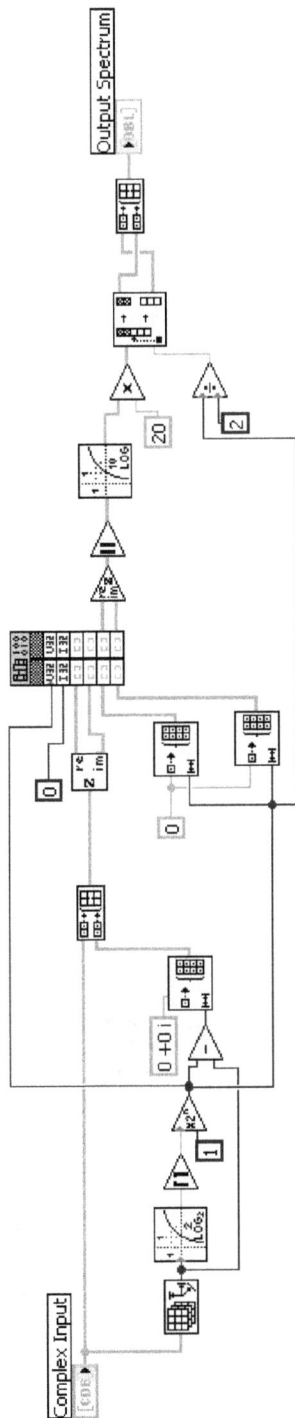

FFT DLL I/O:
Number of Points

Transform Direction: 0 - Forward   1 - Reverse

*Real Input

*Imag Input

*Real Output

*Imag Output

Figure 3.13   Completed FFT DLL call library function node.

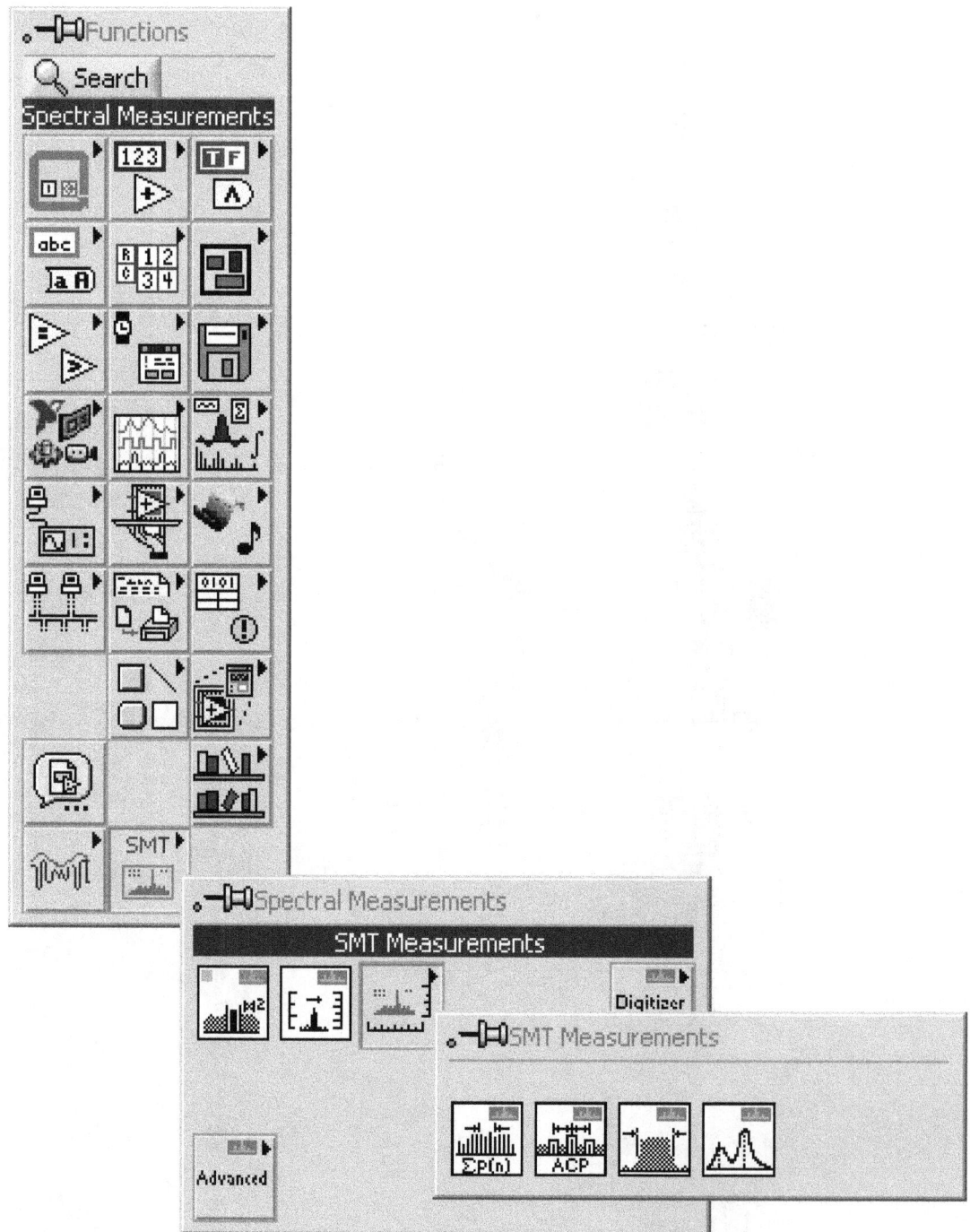

Figure 3.14 Location of spectral measurements toolset.

Figure 3.14 shows the location of the SMT subpalette within the LabVIEW controls. All of these functions are very closely tied to the 5660 hardware and it is for this reason that NI chooses to bundle the SMT and 5660. Browsing around in the SMT subpalette, there are also many low-level VIs that will allow you to configure the NI Tuner (PXI-5600) and the NI Scope (PXI-5620) independently of the RFSA or SMT functions. As with many other NI supplied tools, the SMT functions are designed for the user to easily configure the 5660 hardware for the desired measurements, but the users are also free to build their own applications from the low-level functions.

Summary

This chapter has introduced some of the spectral analysis capabilities built into LabVIEW. In exploring those capabilities, some inherent impairments of the DFT were revealed. They included scalloping losses, spectral leakage, and limited frequency resolution. Since the DFT is a sampling of the continuous Fourier transform, these impairments cannot be eliminated but they can be reduced through the proper use of windowing and sizing of the DFT. We also built a very useful VI called AdvFFT.vi to extend the functionality of the basic DFT routine by forcing the FFT computation and by performing a shift on the output spectrum for display purposes. The final topic of the chapter was the incorporation of DLL libraries into a LabVIEW VI. This functionality adds a new dimension to the processing capabilities of LabVIEW since there are enormous amounts of signal processing libraries available on the Internet. Clearly LabVIEW has some very powerful spectral analysis tools, which we will use extensively in the next few chapters as we cover digital filters and multirate processing on our way to completing this communication system.

References

1. Oppenheim, A. V., R. W. Schafer, and J. R. Buck, *Discrete-Time Signal Processing,* 2d ed., Prentice-Hall, Upper Saddle River, NJ, 1998.
2. Harris, F., "On the Use of Windows for Harmonic Analysis with the Discrete Fourier Transform," *Proceedings of the IEEE,* vol. 66, no. 1, January 1978.
3. Harris, F., *Multirate Digital Signal Processing,* Prentice-Hall, Upper Saddle River, NJ, 2004.

Digital Filters

The topic of digital filters has to be the bread and butter of the digital signal processing (DSP) world. Most types of signal processing will eventually involve the use of a digital filter. Rate conversions, pulse shaping of modulated waveforms, and even examining the discrete Fourier transform (DFT) results all require the use of filters. On top of those applications, we also employ them at will to remove noise from our recovered signals and to eliminate out-of-band spurs before transmission. Like the other DSP topics we have seen, LabVIEW also has a full arsenal of digital filter design tools. This chapter takes an in-depth look at some of the more common filter design tools included with LabVIEW.

4.1 Filter Types

In general, filters can be classified into finite impulse response (FIR) filters and infinite impulse response (IIR) filters. With most of the situations encountered in this book, we will require the use of FIR filters. This type of filter is probably the most common filter in digital communications. The justification for their heavy use comes from the fact that symmetric FIR filters are also classified as linear-phase filters. All filters will not only affect the magnitude of their inputs, but they also impart some phase distortion at the output signal. The term linear phase means that (in the filter passband) the phase influence of the filter coefficients on the input is a linear function. This linear phase distortion winds up causing a constant (and easily removed) time delay. In some speech applications, the filter-phase effects are usually not important because of the way the human ear responds to phase. However, in digital communications, the phase of our signal is very important for correct detection and demodulation. Therefore, we will focus our attention on linear-phase FIR filters and only briefly examine LabVIEW's IIR capabilities.

Figure 4.1 shows the location of LabVIEW's filter palette. The top row of functions contains the common IIR filters and the middle row contains some common FIR filters. The functions shown in this palette are really just interfaces

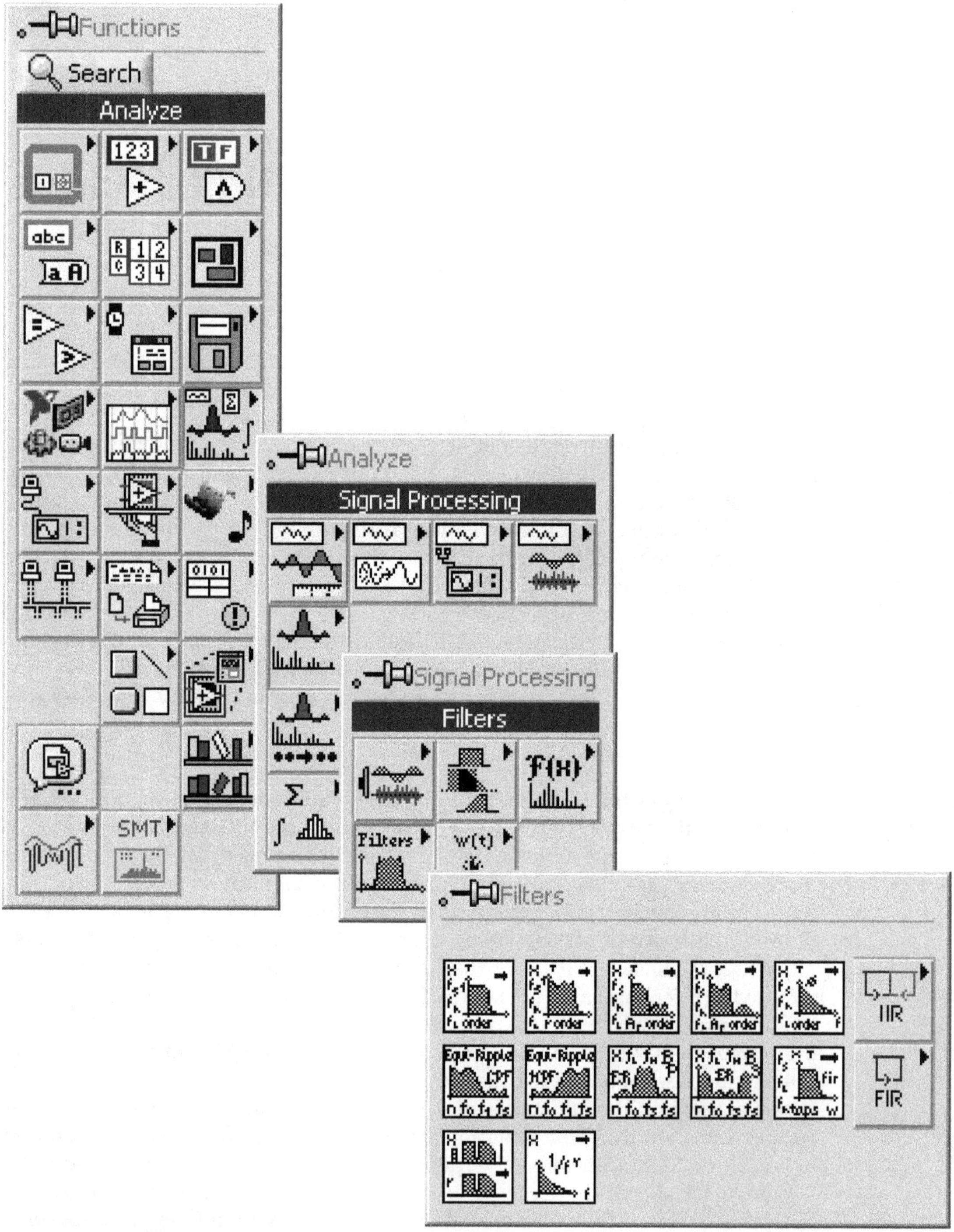

Figure 4.1   Location of LabVIEW filter functions.

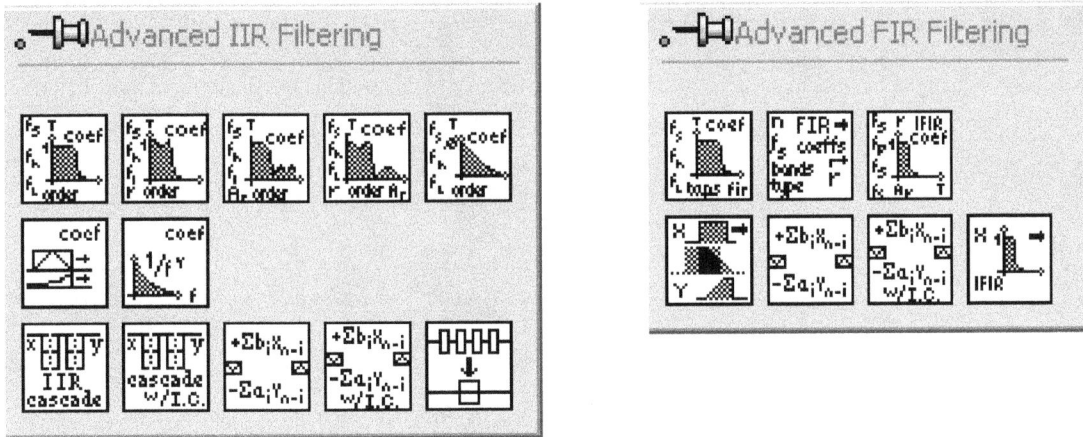

Figure 4.2   Advanced filtering subpalettes.

to the more advanced functions shown in the IIR and FIR subpalettes in Fig. 4.2. Those subpalettes contain the core algorithm virtual instruments (VIs), which compute the filter coefficients for the various filters. This chapter uses the lower-level functions in the advanced subpalettes in order to build some useful filtering VIs.

4.2   FIR Filters

Since FIR filters are simply weighted summations of the present and previous inputs, a very simple FIR filter can be built as in Fig. 4.3. This block diagram shows a simple moving average filter that sums the current and some given number of previous input samples and averages them. This kind of VI can be used to smooth the quantization steps in a waveform captured by an A/D card (this may be known as a moving average filter). This VI has no weightings on the previous input values. If it did have a weight assigned to each coefficient, how would you know what coefficient gets what weight and how many previous inputs would you average? Well, some trial and error, and a look at the filter response might get you close for small numbers of coefficients, but for even fairly short filters this process would be difficult. Besides trial and error, what methods are available for designing this type of filter? FIR digital filter design is not quite as closed form as you might think. However, there are some very good methods for approximating the filter response we are looking for. The next few sections will explain the details of those methods and how to use LabVIEW to design FIR filters from those methods.

4.2.1   FIR filter design by windowing

The windows that we looked at in Chap. 3 had to do with improving the DFT results by filtering the signal at each DFT bin. Since those windows acted as lowpass filters (LPF) centered at each analysis frequency, you might have

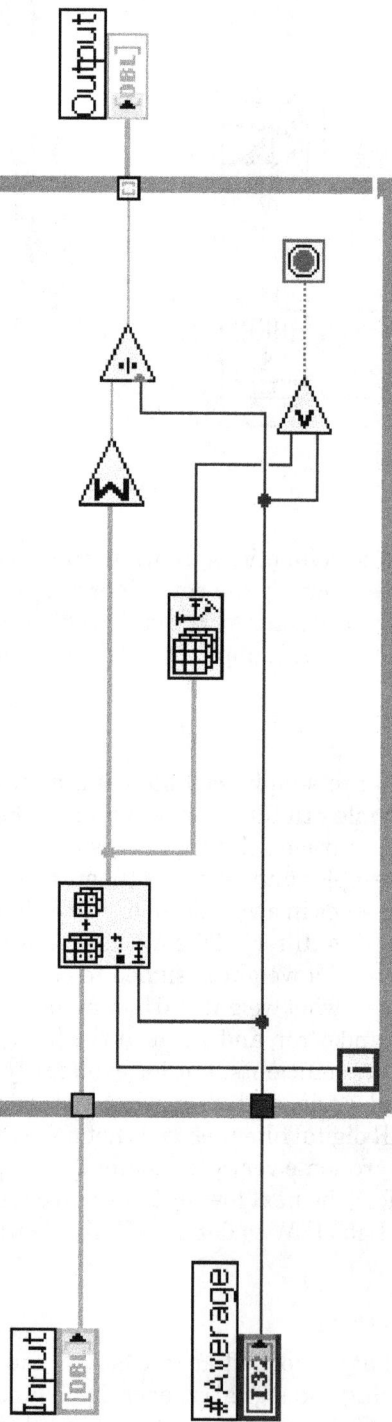

Figure 4.3 Simple moving average filter.

guessed that the same technique is actually useful for processing signals beyond just the DFT. In order to implement a digital filter, it must be both causal and finite. So how do you make a filter finite? That part is easy, just truncate it. In a causal system, nothing can happen before time 0, so to impose causality on a filter, we can simply delay the response by enough time to make it causal. We will easily impose causality by forming a window starting at time 0. Now all you need is something to force these conditions on. For that part, let's just start with the ideal LPF. Figure 4.4 shows both the frequency response of the ideal filter and its time-domain impulse response. You may remember from your college coursework that the sinc function and the rectangular window form a Fourier transform pair. Now of course, in reality the sinc function is infinite in duration and therefore has a very clean ideal frequency response, which is a perfect rectangular window. For display purposes, the sinc was truncated and the frequency response is slightly distorted.

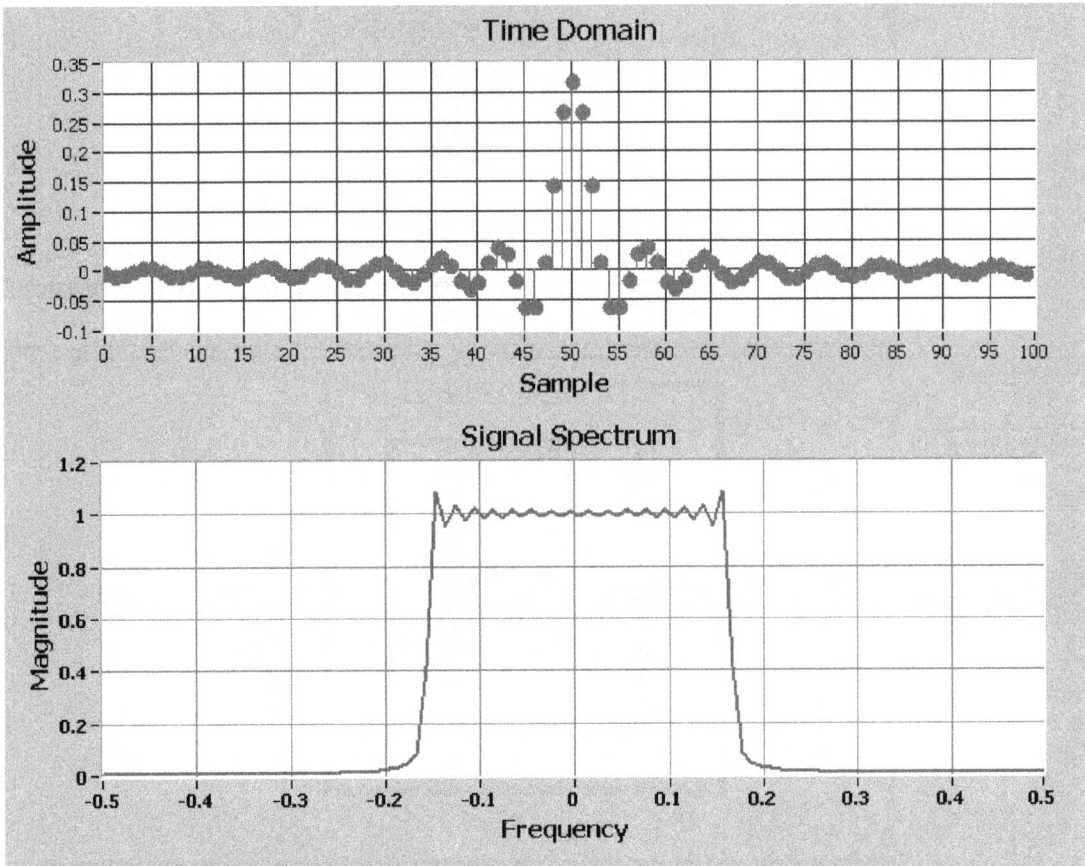

Figure 4.4   Time-domain and frequency-domain plot of sinc function.

The next step in using windowing to design a filter is to choose a window. For the reasons discussed in Chap. 3, the rectangular window has a very narrow main lobe, but has poor side-lobe attenuation and is probably not the best choice for us. The side lobes are the causes of ripple in the pass- and stopbands of the filter and therefore to maintain signal integrity, we would like to minimize the side lobes. So filter design by windowing sounds fairly easy—just choose your ideal frequency response and apply the desired window. Well actually that is only half

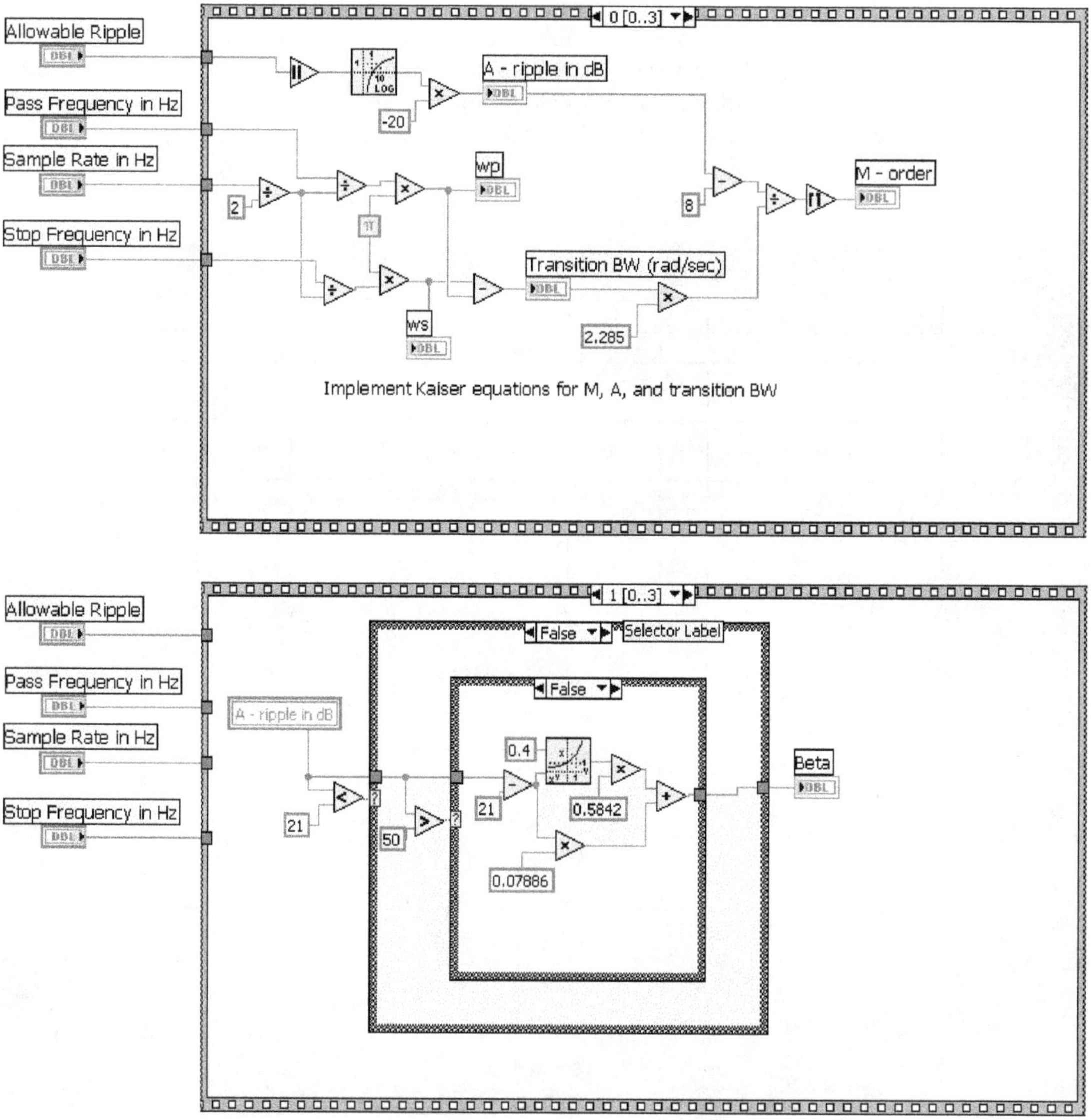

Figure 4.5   KaiserFIR.vi block diagram frames 0 and 1.

the story. It turns out that the resulting filter will have a main-lobe width that is not only dependent on which window function you choose, but also on the time length of that window. Reference [1] contains a table comparing the properties of commonly used windows. So now we have to choose an ideal frequency response, a windowing function, and an appropriate filter length in order to really get the response that we are looking for. This is starting to sound like a lot of variables to try and get everything right where we want it. We may have to go through several iterations of filter length and different windowing functions before we get the filter we are looking for. Here, a man named Kaiser has paved the way for us by developing a window function based on a zeroth-order modified Bessel function of the first kind [1]. The aptly named Kaiser window implements this function and as luck would have it, LabVIEW has included this window as shown in Fig. 3.7. Along with the equation for the near-optimal window, Kaiser also empirically developed some formulas to take most of the guesswork out of the windowing filter design method. These formulas are presented as follows [1].

Kaiser window design formulas

$$\Delta\omega = \omega_s - \omega_p \tag{4.1}$$

$$A = -20\log_{10}\delta \tag{4.2}$$

$$\beta = \begin{cases} 0.1102(A-8.7), & A > 50 \\ 0.5842(A-21)^{0.4} + 0.07886(A-21) & 21 \le A \le 50 \\ 0.0 & A < 21 \end{cases} \tag{4.3}$$

$$M = \frac{A-8}{2.285\Delta\omega} \tag{4.4}$$

Equation (4.1) describes the transition band of the filter in terms of the pass frequency ω_p and the stop frequency ω_s; Eq. (4.2) is the filter attenuation in decibel; Eq. (4.3) computes the shaping parameter β from A; and Eq. (4.4) tells us the recommended filter order M. These formulas are very good approximations and give us an excellent starting point for designing an FIR filter. Now we will use the given formulas to build a VI that will compute these values and design a filter based on our choice of pass frequency, cutoff, ripple, and filter attenuation. The VI will be called KaiserFIR.vi and is shown frame-by-frame in Figs. 4.5 and 4.6.

In this VI, we must first use the Kaiser equations to compute the transition bandwidth, the approximation error A, β (the shape parameter), and M (the filter order). These steps are accomplished in frames 0 and 1. Once we have computed the design values, we can use the LabVIEW function Kaiser-Bessel Window.vi to form the coefficients as shown in frame 2 of Fig. 4.6. You will notice there is an extra step in there to actually form a window, which is one sample shorter than required (filter length should be $M + 1$), then append the zeroth index value to the end of

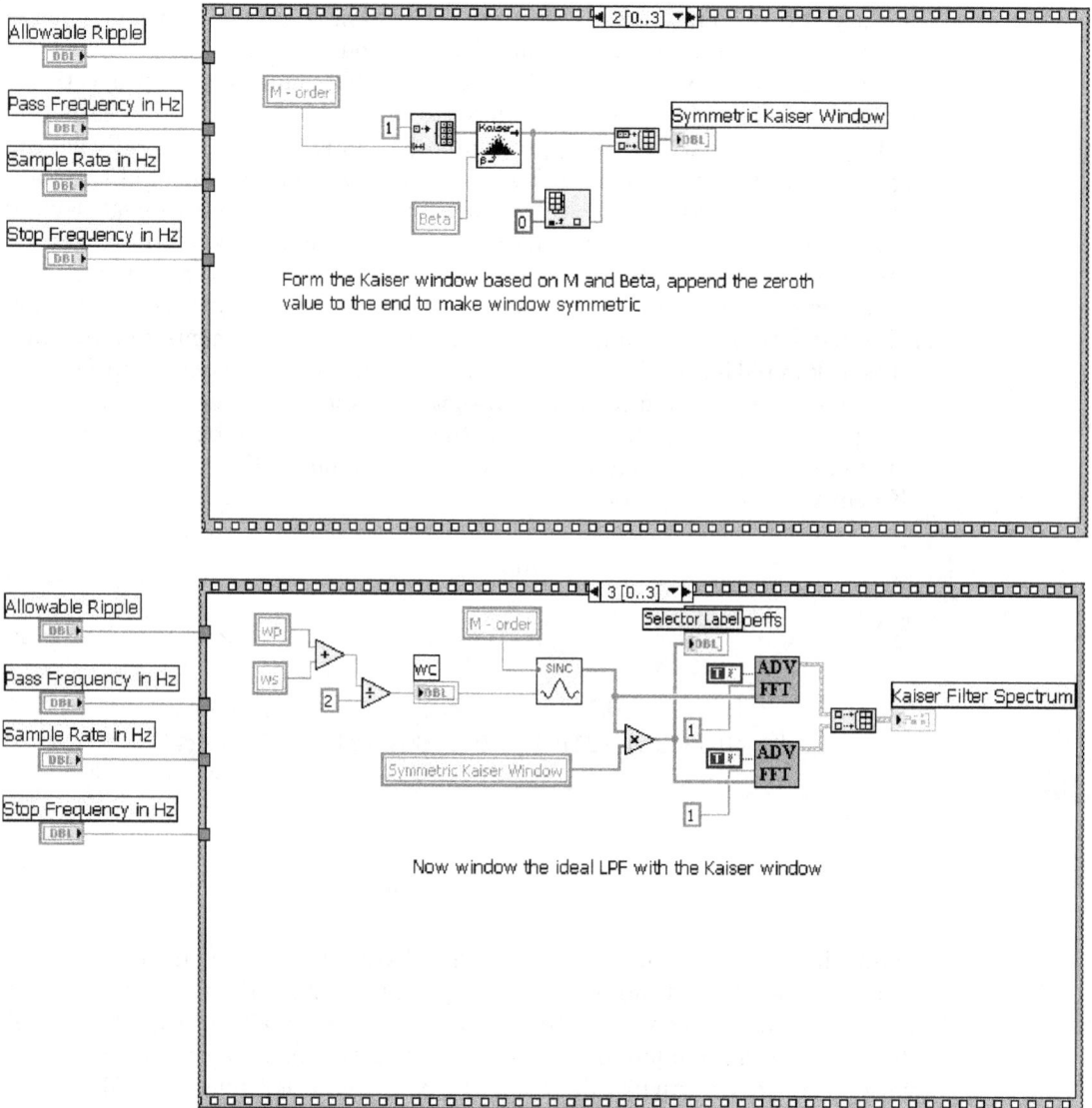

Figure 4.6   KaiserFIR.vi block diagram frames 2 and 3.

the computed window. This is done to make the window symmetric (more on this later). Because the Kaiser-Bessel VI is designed to actually multiply an input signal X by the window coefficients and only give us the windowed output, an array of M 1s is wired to the input along with the β parameter to give the actual window coefficients as output. Once the Kaiser window coefficients are calculated, we have to compute the sinc function. This is done with SincFcn.vi and this VI will be examined more closely in Chap. 6 when we get to generating signals. For now, we will just assume that this function will generate a sinc waveform that is the same length as our Kaiser window. The final step shown in frame 3 is to multiply the Kaiser window by the sinc function in the time domain. The multiplication by a window in time becomes a circular convolution in the frequency domain as we saw back in Chap. 3 and this is what gives us the filter response shown in Fig. 4.7.

Examining the filter response of the KaiserFIR VI, it is immediately obvious that the Kaiser windowed filter is much improved over the rectangular case. It is somewhat difficult to see in this figure, but the rectangular window response has more ripple, which is particularly noticeable in the passband. Remember that ripple in the passband and stopband is caused by the side lobes of the window function. Therefore, because the Kaiser window has lower side lobes than a rectangular window, the ripple is significantly reduced. In addition, we get about 30 dB more attenuation in the stopband with the Kaiser designed filter.

So the Kaiser window gives us a very good approximation to the ideal frequency response. On top of that, the design formulas listed in Eqs. (4.1) to (4.4) take the guesswork out of designing filters by windowing. So what could be better? Well for one thing the ripple in the passband and the stopband are based on the same ripple value δ. Filter designers typically would like to allow different ripple in the passband than in the stopband. That means that you could trade passband ripple for stopband ripple and meet the same filter requirements with a lower-order filter [1]. In the next section, we will look at an algorithm that computes filter coefficients based on the principle of relaxing the ripple requirement in some bands to get improved performance in other bands.

4.2.2 Equiripple FIR filters

As mentioned in the previous section, it is possible to trade ripple in the filter passband for improved ripple in the stopband. By performing this trade, a shorter length filter can be designed based on the same parameters [1]. An efficient algorithm for performing this trade was developed by Parks and McClellan and is therefore known as the Parks-McClellan algorithm. The details of this algorithm can be found in [1] as well as almost any other good DSP book. For this discussion, it is not important for us to know the exact details of the Parks-McClellan algorithm. The algorithm has become a de facto standard for FIR filters and is commonly bundled with most filter design software tools, including LabVIEW. This algorithm is based on the minimization of the mean square error and therefore results in the *best* filter that can be designed to *just* meet the given input criteria. One consequence of this type of filter design is equiripple response in the passband and the stopband. As with the Kaiser window method, the filter order is empirically derived to be [1].

$$M = \frac{-10\log_{10}(\delta_1\delta_2)-13}{2.324\Delta\omega} \tag{4.5}$$

where δ_1 and δ_2 are the ripple constraints in the two bands and $\Delta\omega$ is the transition bandwidth.

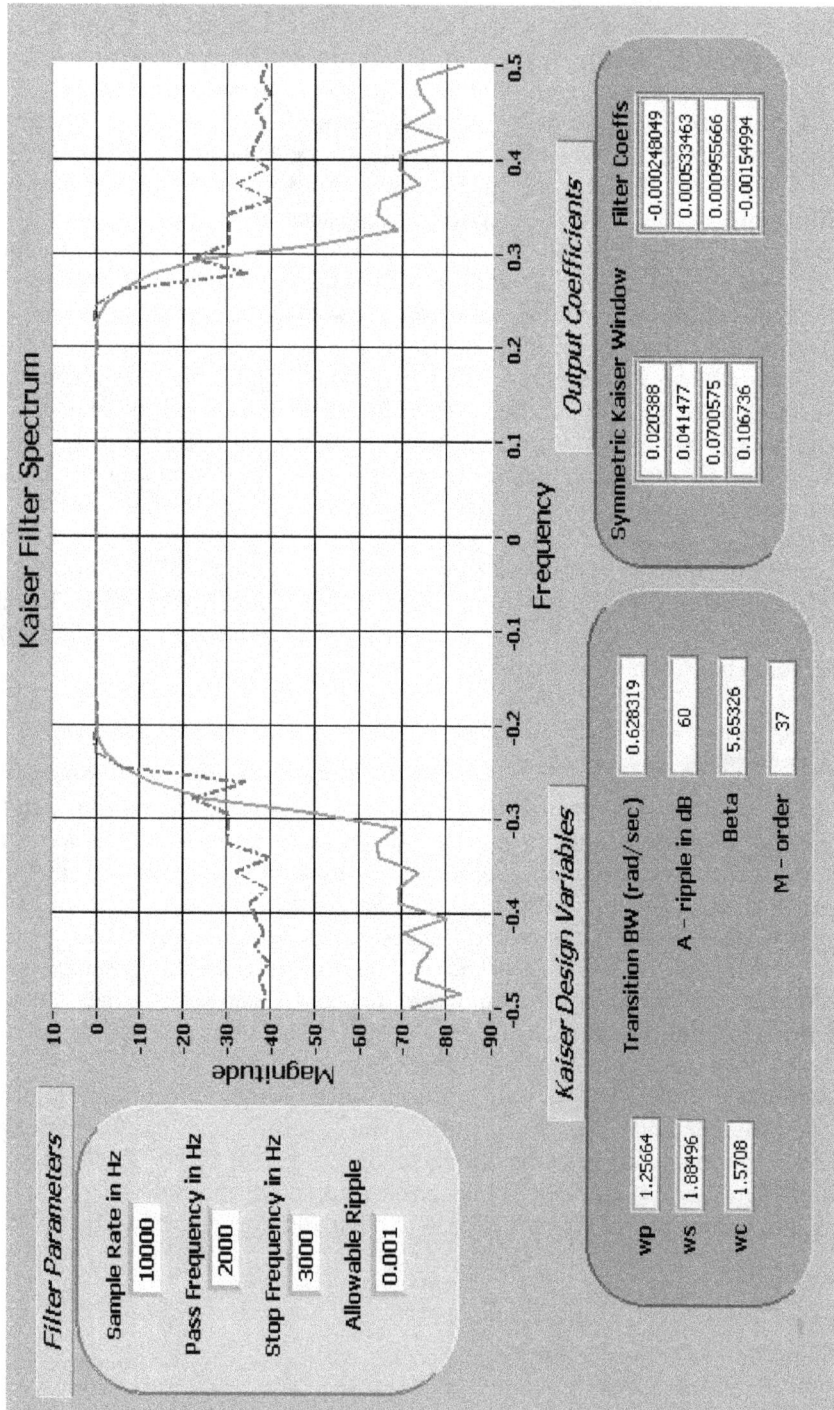

Figure 4.7 KaiserFIR.vi front panel; Kaiser windowed response show as dark solid line, rectangle window response shown as light dashed line.

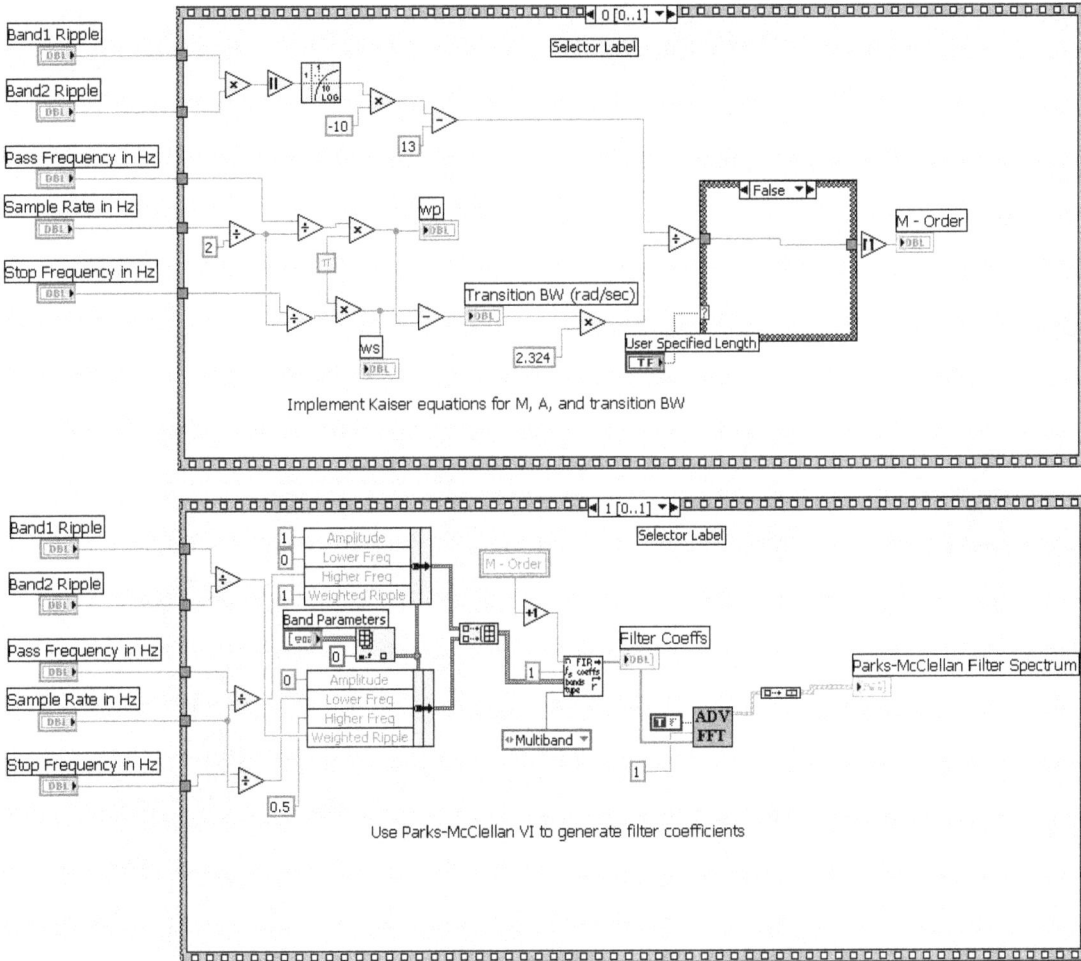

Figure 4.8 MPR.vi block diagram. This VI designs a LPF based on Parks-McClellan algorithm.

Using the formula for filter order in Eq. (4.5) and the LabVIEW VI Parks-McClellan.vi, the block diagram in Fig. 4.8 can be built. The first frame simply computes the value of M from the equation shown in Eq. (4.5). There is also a case selector to allow the user to specify the filter length rather than be forced to use the computed value. Since the Parks-McClellan algorithm typically designs the shortest filter that will just meet the given criteria, allowing the user to control the filter length lends a little more flexibility to the program. This is also particularly useful because some authors suggest that the empirically derived Eq. (4.5) frequently underestimates the required filter order. The second frame is where the Parks-McClellan algorithm is called. The inputs to the algorithm are designed in such a way as to enable any type of filter to be designed (lowpass, bandpass, or highpass) by entering parameters for each band. In this case, we are designing a lowpass filter that has only two bands. Notice also that the weighted ripple in the

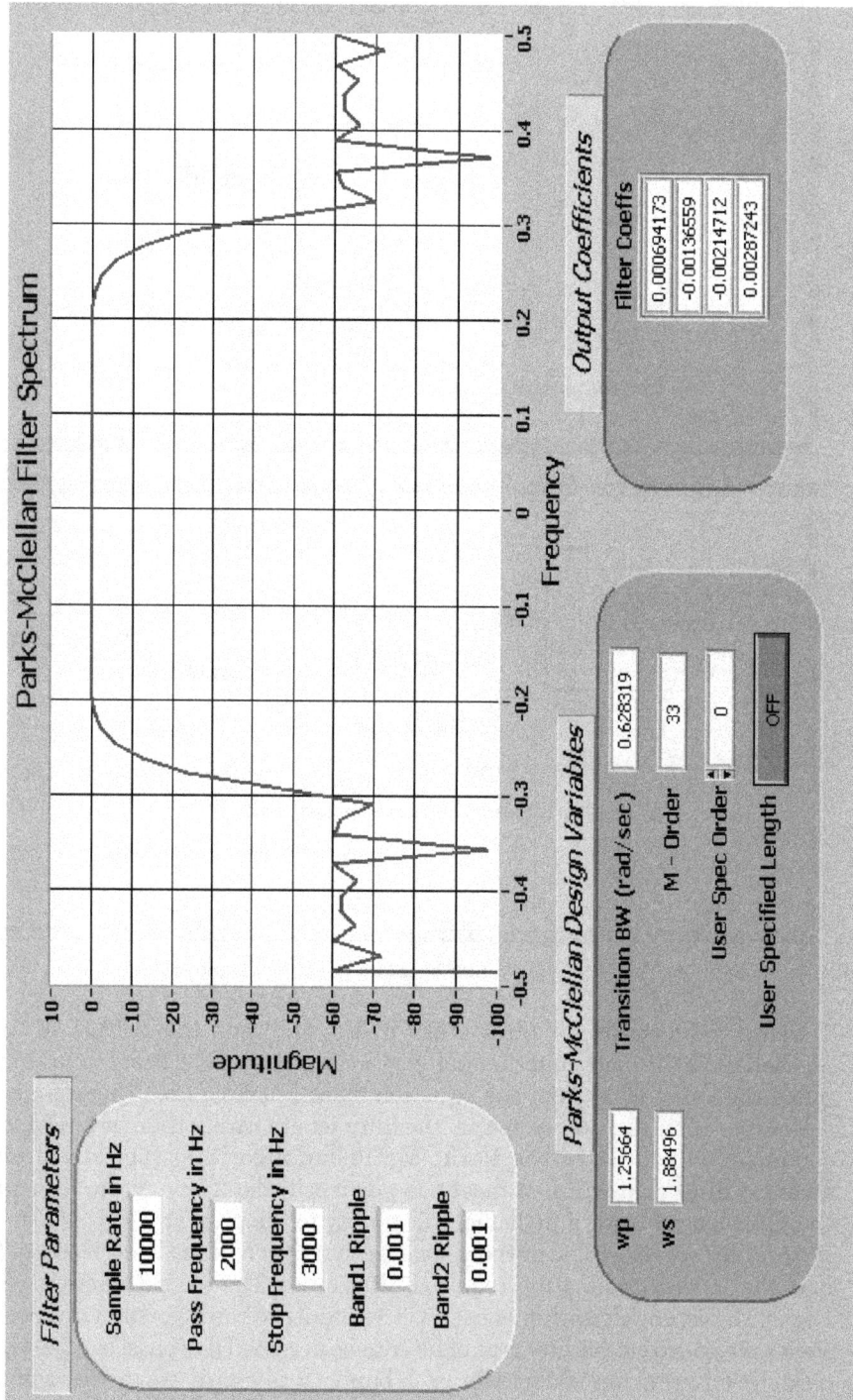

Figure 4.9 MPR.vi front panel. Spectrum of LPF shown.

passband is a constant 1 while the weighted ripple in the stopband is the ratio of passband ripple to stopband ripple.

Figure 4.9 shows the spectrum of a lowpass filter designed using our MPR.vi. As we expected, since the algorithm produces the best filter that just meets the given filter parameters, the Parks-McClellan function produces a shorter filter (34 taps versus 38) with the same parameters as the Kaiser-designed filter shown previously in Fig. 4.7. Note also that in this example, the passband and stopband ripple were chosen to be the same. For an even greater reduction in filter length, we could specify a lower passband ripple while leaving the stopband ripple the same. Also you can clearly see the equiripple characteristic in the stopband.

4.3 IIR Filters

I mentioned before that in digital communications we use a lot of linear-phase FIR filters. While that is true, we certainly cannot ignore IIR filters altogether. IIR filters are recursive and do not exhibit linear phase. While this character-istic is not desirable for most filtering in digital communications, there may be cases where you want to get more attenuation with a lower-order filter. In many instances, a lower-order IIR filter can match or beat the filter perform-ance of an FIR filter, and that can translate into speed improvements for the filtering operation. And of course LabVIEW has several tools for designing IIR filters as shown in the top row of the filters subpalette in Fig. 4.1. These func-tions include the Butterworth, Chebychev, and Elliptic filters. Similar to the FIR filters, we will use the built-in Butterworth Coefficients.vi function to gen-erate a set of IIR filter coefficients. Again we will first want to know the order of the filter that is required. For a Butterworth approximation, the filter order is given by [2] as

$$N = \frac{\log\left[\left(\frac{1}{\delta_2}\right) - 1\right]}{2\log\left(\frac{\Omega_s}{\Omega_c}\right)} \tag{4.6}$$

The VI, ButterworthLPF.vi is shown in Fig. 4.10. If we were looking for the shortest filter for the job, we probably should have chosen the Elliptic Coefficients.vi instead of the Butterworth. However, the elliptic filter order is given by a complex computation that is generally looked up in a table, whereas the Butterworth design has the nice closed form expression shown in Eq. (4.6). The Butterworth coefficient function actually gives us two sets of filter coeffi-cients, the forward and reverse coefficients. These coefficients relate to the over-all filter response as follows [2]:

$$h[n] = \frac{\sum_{k=0}^{M} b_k[n]}{1 + \sum_{k=1}^{M} a_k[n]} \tag{4.7}$$

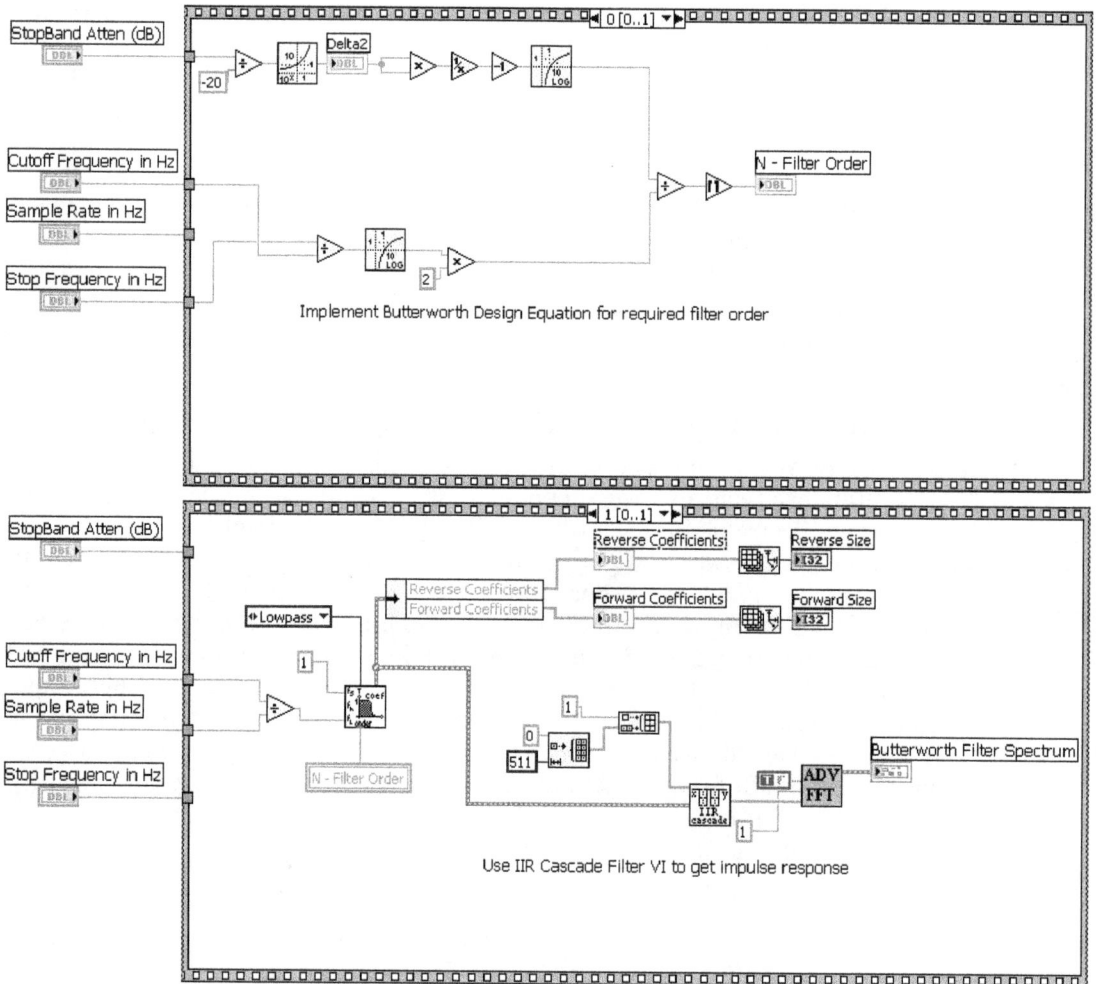

Figure 4.10   ButterworthLPF.vi block diagram.

where the b_k's are the forward coefficients and the a_k's are the reverse coefficients. The block diagram in Fig. 4.10 uses the IIR Cascade.vi to extract the impulse response from the two sets of filter coefficients (a_k's and b_k's). You can see in the block diagram that an impulse function was passed as the input to IIR Cascade.vi in order to get the impulse response as output. Finally we see the plot of the impulse response of the Butterworth designed IIR filter in Fig. 4.11. You can see that even for fairly low filter orders the transition band of the IIR filter is steep and the attenuation in the stopband is large.

4.4   Comparing IIR and FIR Filters

From the discussion in Sec. 4.3, it seems like the IIR filter would be a much better choice for a filter than the FIR we saw earlier. However, there are some

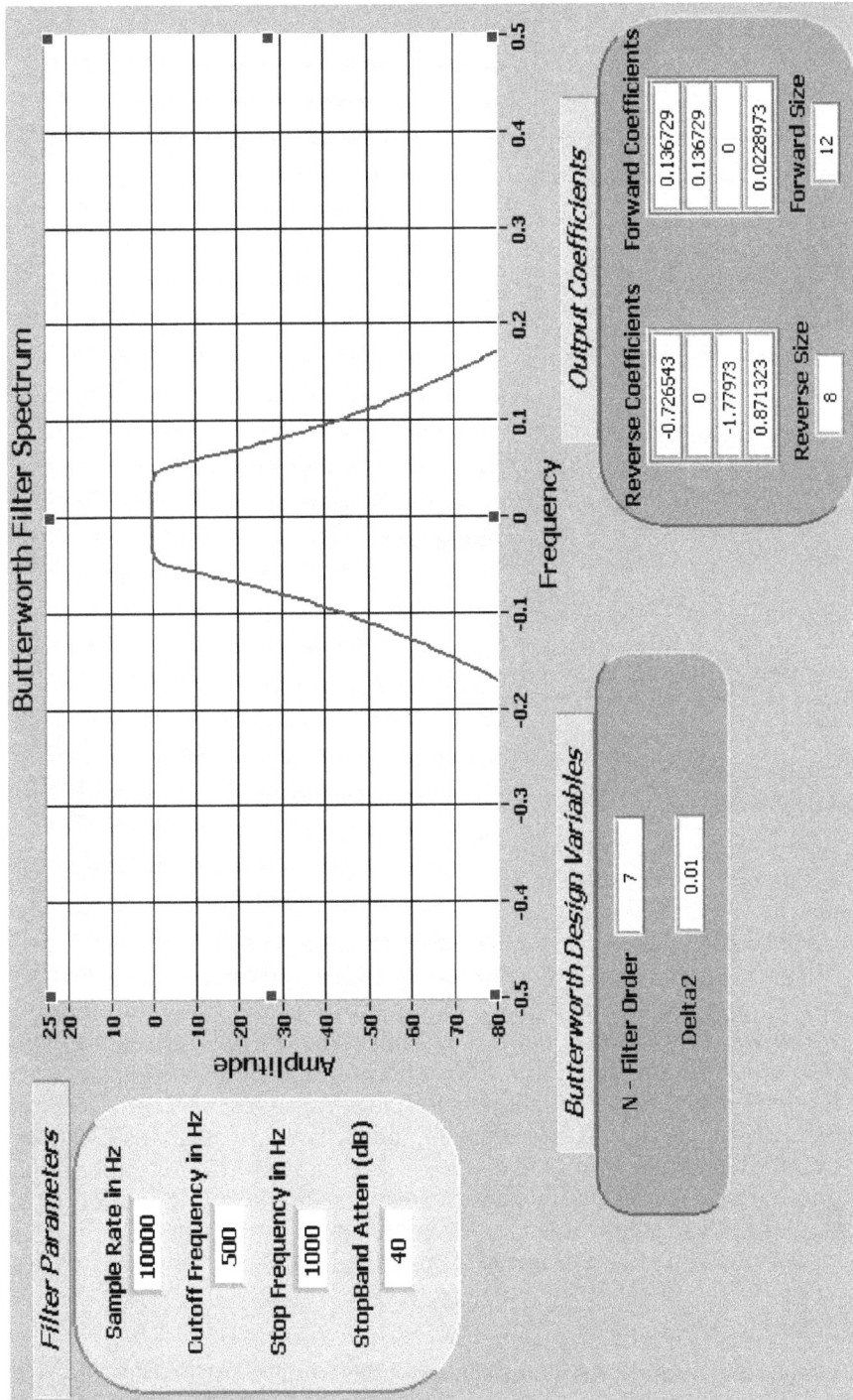

Figure 4.11 ButterworthLPF.vi front panel.

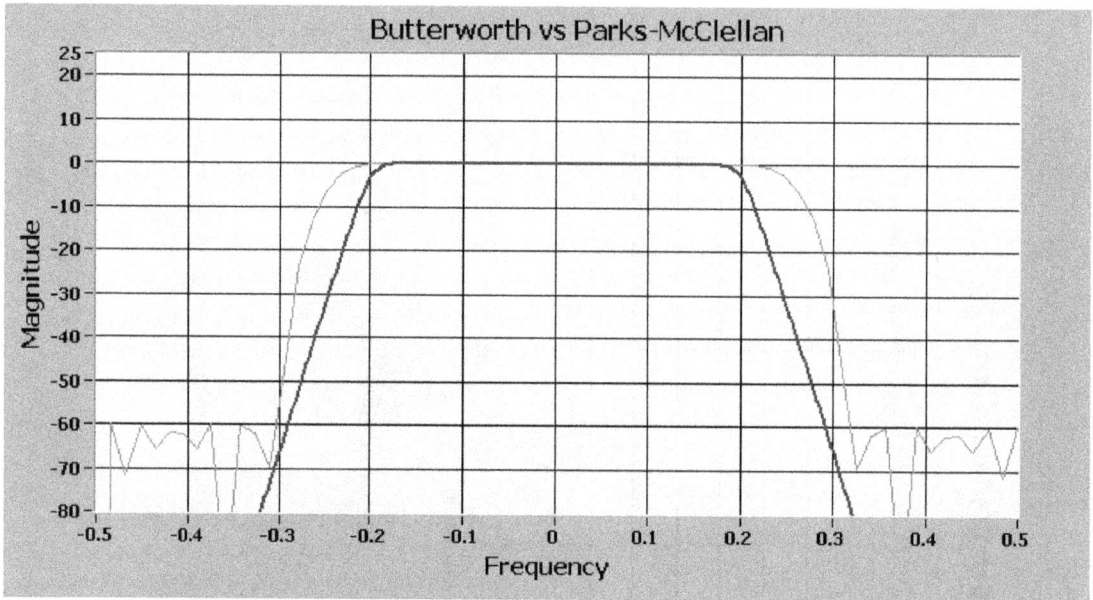

Figure 4.12   Magnitude of Butterworth versus Parks-McClellan.

subtleties that should be examined before we decide which filter to use for a particular application. The next sections compare both the magnitude and phase of these two common filter types.

4.4.1 IIR versus FIR magnitude

Figure 4.12 shows the magnitude comparison of a Butterworth filter (darker plot) designed with the same cutoff and stop frequency as the Parks-McClellan (lighter plot) designed filter in Sec. 4.2.2. The ButterworthLPF.vi function produced a twelfth-order filter which has 13 denominator coefficients and 18 numerator coefficients for a total of 31 multiplications per filter output. The Parks-McClellan filter has 33 taps and thus requires 34 multiplications per filter output. While in this case there is only a difference of 3 multiplications per filter output, we can see from the figure that the Butterworth filter parameters could be relaxed slightly, allowing a lower filter order and fewer multiplications—giving us an even faster filtering operation. If the IIR filter is faster, then why did we like the FIR filters so much again? Remember in digital communications, there is often information in the phase of our signal so we certainly do not want to mess up that information by distorting the phase in a way that we cannot even recover the signal anymore. Let us take a look at how the phase of our signal is affected by these two filters.

4.4.2 Effects of filter-phase response

Figure 4.13 compares the wrapped phase of the Butterworth and the Parks-McClellan designed lowpass filters. The top plot shows that the FIR filter does indeed have

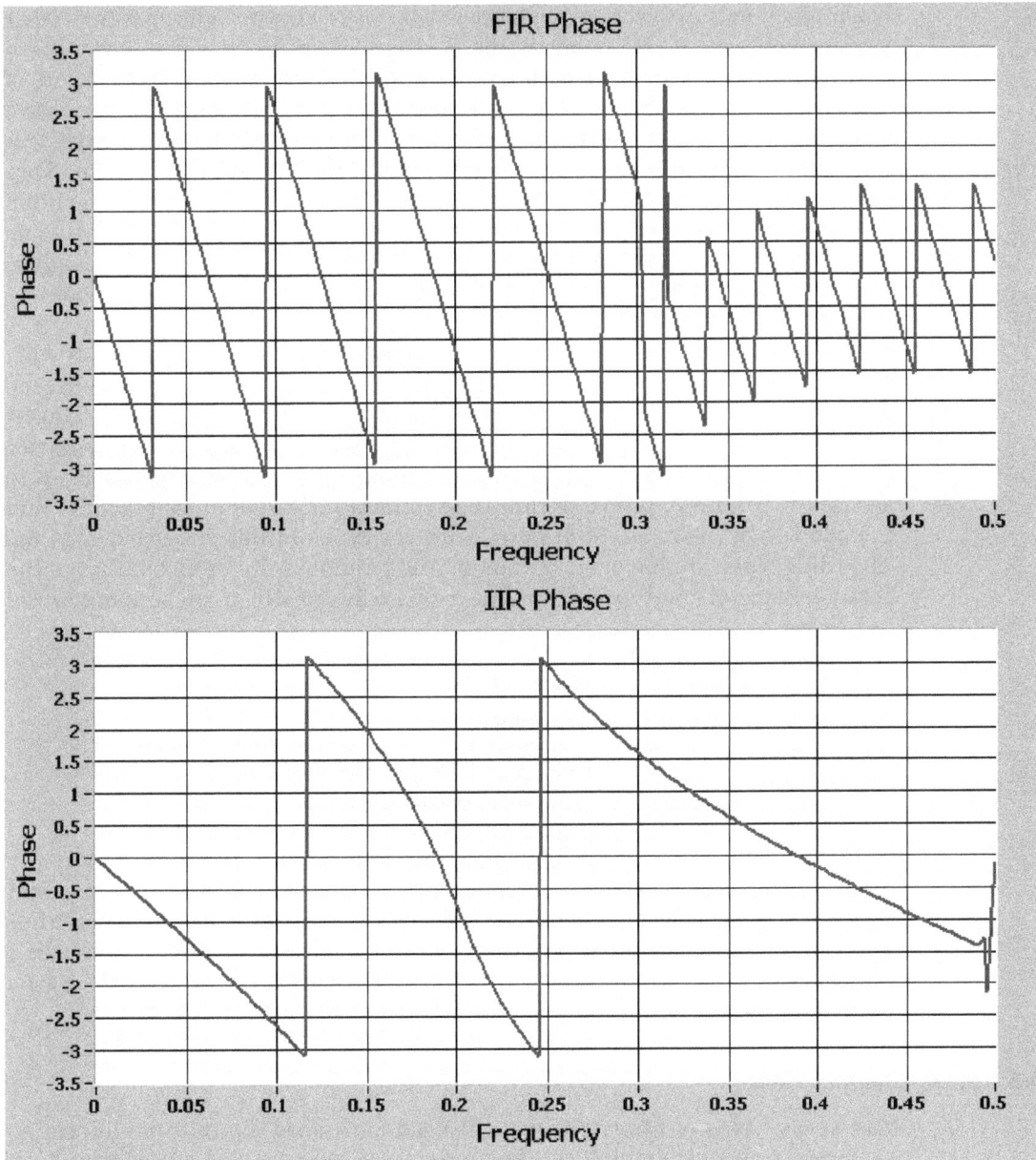

Figure 4.13 Phase of FIR linear phase (top) versus IIR filter.

linear phase in the passband (evidenced by the piecewise straight lines from 0 out to the cutoff at $0.3f_s$). Conversely the bottom plot shows that the Butterworth filter's phase is nonlinear in both the passband and stopband. The real figure of merit related to the filter's phase is group delay. This is simply the derivative (or slope) of the phase. In the case of a linear-phase filter, the group delay is a constant over the passband. This means that all frequencies in the passband are delayed by the same constant amount of samples as they pass through the filter. An IIR filter on the other hand will have a nonconstant group delay and some frequencies will pass with different delays than other frequencies. This leads to a temporal scrambling of the signal that may now be impossible to demodulate correctly.

Linear-phase FIR filters have a well-defined group delay that is simply $M/2$, where M is the order of the filter. Depending on whether M is odd or even, the delay will either be an integer value or it will be an integer plus $\frac{1}{2}$ [1]. Thus it is possible to create half-sample delay filters, which are used in some lattice filter implementations. When designing a linear-phase filter, it is important to choose the filter order M such that the symmetry of the impulse response and the oddness or evenness of M agree with the type of filter you are designing. The filter types are lowpass, highpass, and bandpass. Reference [1] gives the characteristics of four types of FIR linear-phase filters, which can be summarized as follows:

Type I: symmetric impulse response, M even

Type II: symmetric impulse response, M odd

Type III: antisymmetric impulse response, M even

Type IV: antisymmetric impulse response, M odd

The reason that these characteristics are important to the type of filter you are designing is that each of the types has a magnitude response that may suit only one of the previously mentioned filter types. A type II filter for instance has a zero at π, making it an appropriate choice for a lowpass filter, but not for a highpass. Oppenheim and Schafer [1] have plotted example responses for each of the FIR filter types.

4.5 Pulse-Shaping Filter

One special type of filter that we will need for a good digital communication system is a pulse-shaping filter. The term pulse refers to the symbol waveform transmitted every symbol time, T_{sym}. You have probably read somewhere that the radio frequency (RF) channel can be modeled as a linear time invariant filter. Thus, the channel has the effect of spreading out (think convolution) the transmitted symbol waveform. The time spreading smears the transmitted waveform into adjacent symbol time slots. This phenomenon is known as intersymbol interference or ISI. The way to combat ISI is to run the baseband waveform (pulse) through a shaping filter before transmission. Of course now we have to figure out what filter to use for this shaping. Nyquist has developed a criterion for choosing a filter that is guaranteed to have zero ISI [3]. The criterion can be stated as [3]:

$$h(nT_{\text{sym}}) = \begin{cases} 0 & \text{for} \quad n \neq 0 \\ 1 & \text{for} \quad n = 0 \end{cases} \tag{4.8}$$

In words, Eq. (4.8) says that the impulse response for a filter having zero ISI will have an amplitude of 1 at the zeroth multiple of the symbol time and 0 at all other multiples of the symbol time. One such filter is the raised cosine filter, which has the frequency response shown in [4].

$$H(f) = \begin{cases} 1 & \text{for} \quad |f| < 2W_0 - W \\ \cos^2\left(\dfrac{\pi}{4} \dfrac{|f| + W - 2W_0}{W - W_0} \right) & \text{for} \quad 2W_0 - W < |f| < W \\ 0 & \text{for} \quad |f| > W \end{cases} \tag{4.9}$$

From Eq. (4.9), the block diagram shown in Fig. 4.14 can be built. This VI is called NyquistPulse.vi and generates a time-domain pulse shaping window from the frequency response $H(f)$ shown previously. There are a few things to notice about this VI. One is there is a case selector to choose between a raised cosine response and a root-raised cosine response. This selection is necessary because in many cases, a root-raised cosine window will be applied at the transmitter, and the corresponding root-raised cosine window is applied at the receiver. The split window application (root-raised at both ends) gives the system an overall raised cosine response. The signal bandwidth W and the Nyquist minimum bandwidth W_0 are calculated using [2]:

$$W_0 = \frac{R_S}{2}$$

$$W = \frac{1}{2}(1 + \alpha)R_S$$

where R_S is the symbol rate and alpha is the roll-off or excess bandwidth factor.

The left part of Fig. 4.14 is simply building the frequency variable f and computing the required bandwidth W. The frequency array is formed in two separate pieces from 0 to $f_s/2$ and from $-f_s/2$ to 0—then concatenated together into a single frequency array. The loop contains some case structures to compute $H(f)$ based on the conditions given in Eq. (4.9). On the right, we now have the frequency response of this pulse-shaping filter (with choice of root-raised or simply raised cosine) and we use the inverse Fourier transform to calculate the time-domain impulse response. The LabVIEW inverse Fourier transform expects the input to be ordered in frequency from 0 to 2π, where 0 to π are the positive frequencies and π to 2π are the negative frequencies. Now it should be clear why

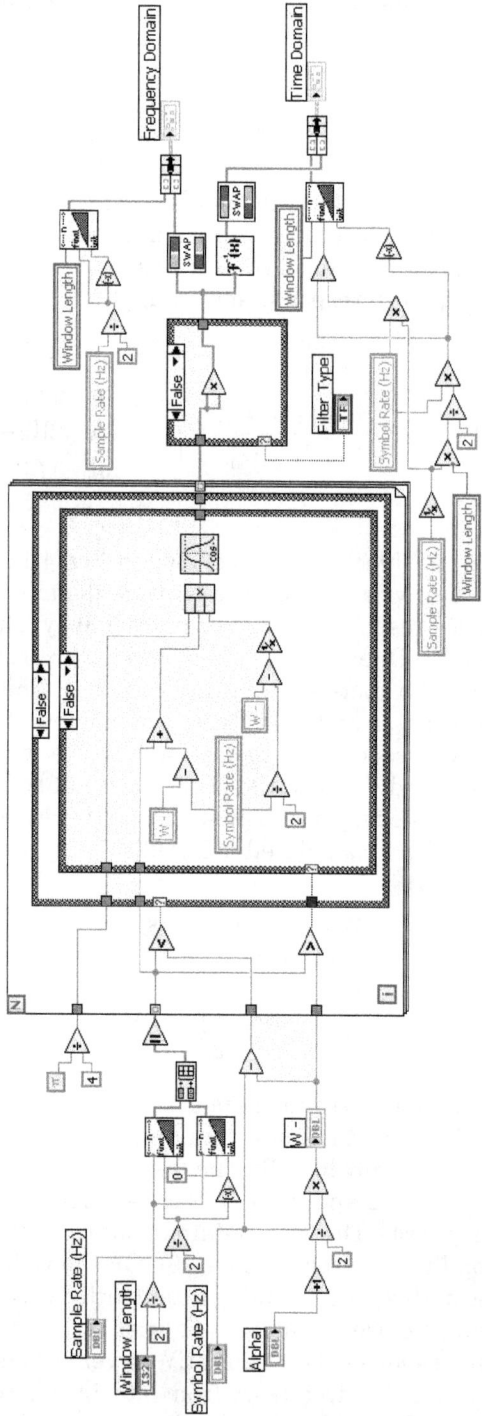

Figure 4.14 NyquistPulse.vi block diagram.

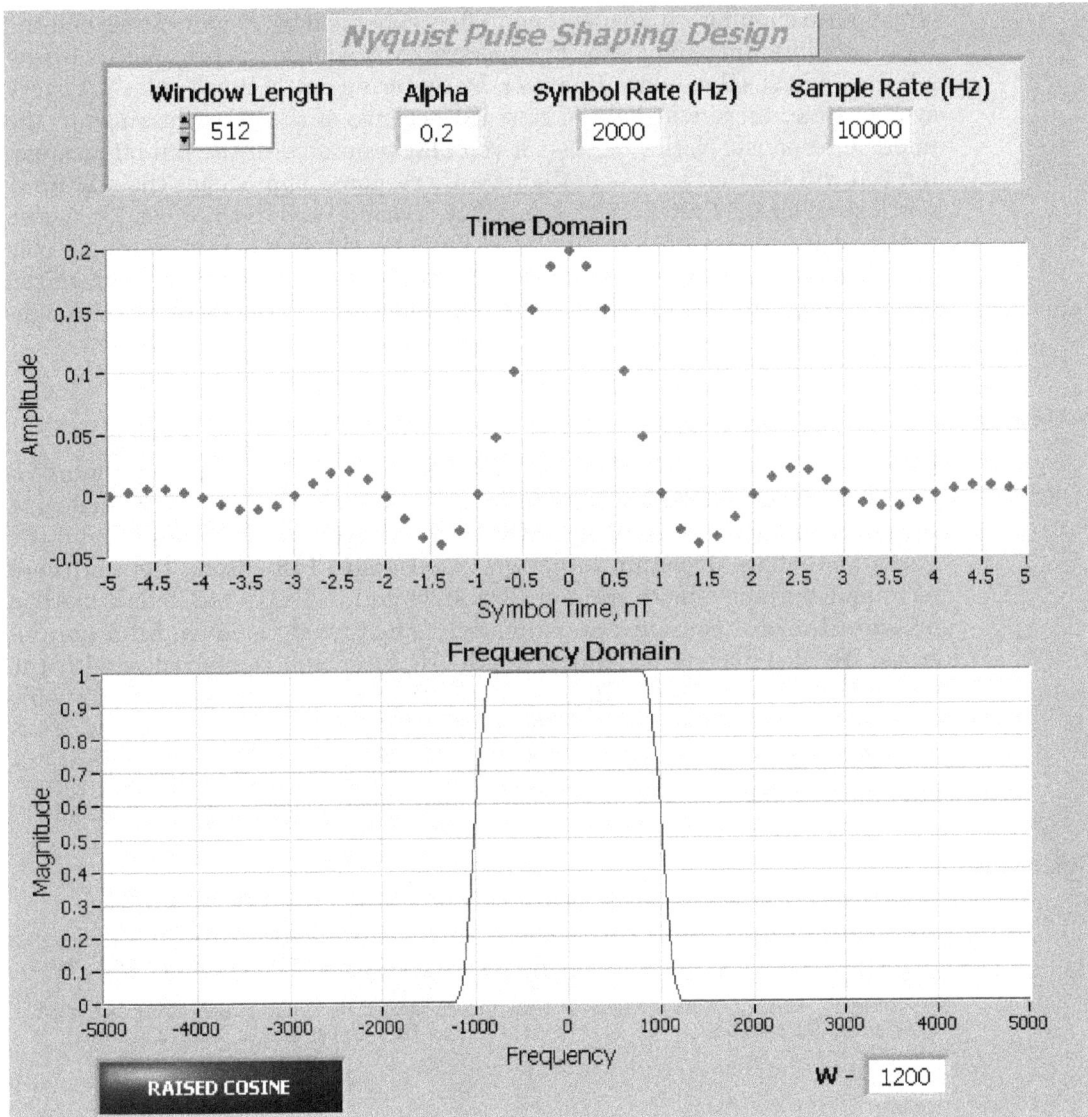

Figure 4.15 NyquistPulse.vi front panel.

the frequency variable was formed in those two separate pieces. Finally, for the reason just mentioned, we use ArraySwap.vi to split the frequency domain and time domain data in half and swap the order to arrive at the display order we are more used to seeing in Fig. 4.15.

The time-domain impulse response at the top of Fig. 4.15 shows that at multiples of the symbol time T_{sym} the amplitude is indeed zero, which satisfies Eq. (4.8). Looking at the frequency domain, we can see the effect of the excess bandwidth of the pulse shaping window, the signal has expanded by 20 percent beyond the Nyquist minimum bandwidth. The spectral expansion is the price for requiring a finite and causal window. However, by centering one of these pulses at every symbol time, there will still be zero ISI because of the zero crossings in the impulse response. Notice too that if you play around with the roll-off parameter, you will see that as $\alpha \to 0$, the impulse response approaches the sinc function (except for the finiteness and causality of course) and the frequency response is almost the ideal LPF. The final note on pulse shaping is that there are also other shaping functions that meet the Nyquist criterion, but the root-raised cosine is very common in digital communications and is the one we use throughout this book.

Summary

We have now covered the core LabVIEW capabilities for designing digital filters. This chapter drew on those tools to build some VI applications to design both FIR and IIR filters. We used both the Kaiser window method and the Parks-McClellan algorithm to design linear-phase FIR filters. Both of these techniques produce near optimal filters, while the Parks-McClellan method produces the best possible approximation to just meet the given filter parameters. We also designed a Butterworth IIR filter and compared it with the Parks-McClellan designed filter in both magnitude and phase. Finally, this chapter introduced the concept of Nyquist pulse shaping and we saw how to formulate a raised cosine window for reducing ISI in a digital communications system.

References

1. Oppenheim, A. V., R. W. Schafer, and J. R. Buck, *Discrete-Time Signal Processing,* 2d ed., Prentice-Hall, Upper Saddle River, NJ, 1998.
2. Proakis, J. G., and D. G. Manolakis, *Digital Signal Processing, Principles, Algorithms, and Applications,* 3d ed., Prentice-Hall, Upper Saddle River, NJ, 1996.
3. Harris, F., *Multirate Digital Signal Processing,* Prentice-Hall, Upper Saddle River, NJ, 2004.
4. Sklar, B., *Digital Communications,* 2d ed., Prentice-Hall, Upper Saddle River, NJ, 2001.

Multirate Signal Processing in LabVIEW

Now we are getting into some really interesting and complex material. Often in the world of digital communications, we run into the problem of having digital data sampled at one rate that we would like to change to another rate. Perhaps it was a hardware limitation or some processing requirement that forced the data to be sampled at a certain rate, but now you need to operate at a different rate. This is where multirate signal processing comes in. Using these techniques you can upsample to a higher rate, downsample to a lower rate or a combination of the two. Typically, the upsampling and downsampling occur at integer multiples of the starting sample rate and by combining the two operations, we can arrive at any noninteger multiple of the original rate. We will look at some fairly simple operations to change the sample rate of some signals and we will also see that there are some very elegant methods for performing efficient rate changing operations.

5.1 Upsampling

The most straightforward way to upsample data to a higher rate is to zero-stuff the signal, meaning put a zero between each sample to increase the number of samples captured in a given amount of time. The process of zero-stuffing adds no spectral content to the original signal, so this seems like a very nice clean way to increase the sample rate of a given digital signal. Now let us build a virtual instrument (VI) to upsample a signal. Figure 5.1 shows the block diagram of Upsample.vi. This VI initializes an array of $I - 1$ zeros and inserts that array between each original input sample.

Figure 5.2 pretty much sums up what happens when you zero-stuff. If we use the filter we designed in Chap. 4 with the KaiserFIR.vi function, we should already know what the spectrum will look like. I said before that no spectral information was added and that is still true. What did change, however, was the

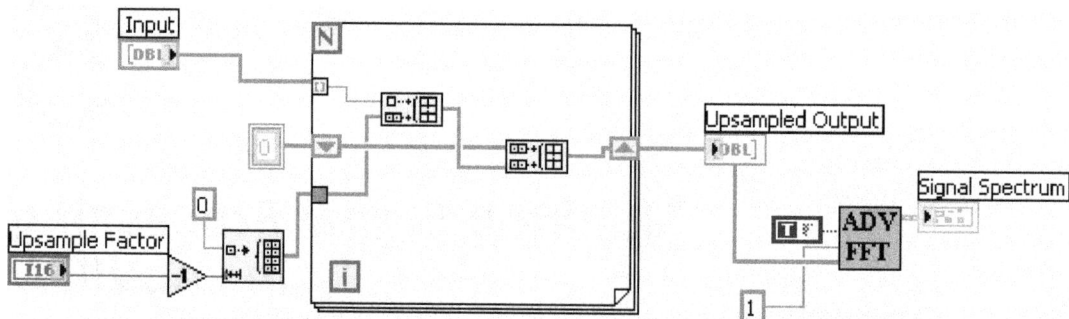

Upsample.vi stuffs zeros between each sample of the input signal.

Figure 5.1   Upsample.vi block diagram.

digital frequency location of the spectral replications. When dealing with digital signals it is important to remember that digital frequency depends on the sample rate. Now that we have zero-stuffed our input signal, we have three times as many samples in the same time period. Effectively, we are viewing a larger snapshot of the original signal's periodic spectrum. Before upsampling, we could only see

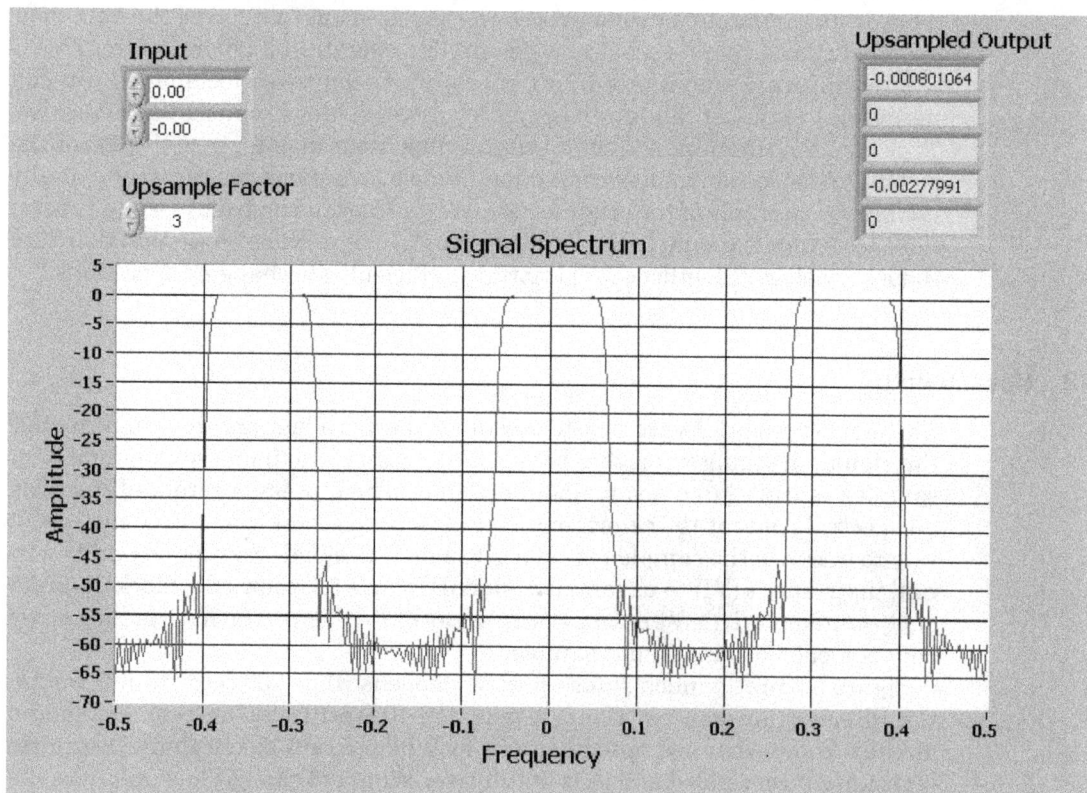

Figure 5.2   Upsample.vi front panel. Three spectral replications are shown when upsampling by 3.

from $-f_s/2$ to $f_s/2$. Now that f_s got bigger while the spectral content stayed the same, we can see three times as much of the original periodic spectrum.

There may be some interesting applications where you want to keep the spectral images from the upsampling but in most applications we are going to have to filter this signal to remove those unwanted spectral replications. Before we look at any filtering of resampled signals, let us take a look at the reverse operation, downsampling.

5.2 Downsampling

If you can think of upsampling as compressing more of the periodic spectrum into the space from $-f_s/2$ to $f_s/2$, then you can imagine what happens when you downsample. The spectrum spreads out like you are zooming in on some portion of the original spectrum and causing it to fill the space from $-f_s/2$ to $f_s/2$. The danger with this operation is that the "zoomed" spectrum can overrun the $f_s/2$ boundaries. The overrun causes aliasing and you may not be able to recover the portion of the signal you are interested in. Also keep in mind that you are throwing away pieces of your original signal that contribute to the total power in your original spectrum. That means that you will lose signal power proportional to the downsampling factor in the operation.

Figure 5.3 shows the spectrum of a downsampled signal along with the spectrum of the original signal. Notice that the signal was downsampled by a factor of 2 and consequently the signal bandwidth has doubled while the amplitude was cut in half ($1/2 \sim -6$ dB). If we were to increase the downsample factor in Fig. 5.3 any more, to say, 3—then we would start to see aliasing. That is because the original signal has spectral information all the way out to 2000 Hz. If we expand the signal by 3, then components from 2000 Hz move out to 6000 Hz, and we have caused aliasing by going beyond the half sample rate of 5000 Hz (remember from Chap. 4 that the sample rate for these filter coefficients was 10,000 Hz).

From this example it looks as if we can avoid aliasing in the downsampling-by-2 operation as long as our original signal bandwidth is always less than $1/2$ of the half sample rate or more directly $1/4$ of the original sample rate. But sometimes our original signal may include noise that gives us significant spectral content at frequencies greater than the $1/4$ sample rate. When we downsample that noisy signal, the noise may alias and distort the resultant signal. Of course this phenomenon is not limited to sample rate reductions by a factor of 2. For an arbitrary sample rate reduction by M, the signal must be *first* band-limited to $f_s/2M$ [1]. That is why it is necessary to run the signal through an antialias filter before the downsample operation. Now that we have seen an overview of the upsampling and downsampling operations, let us complete the resampling operation by looking at the filtering process.

5.3 Resampling Filters

As we saw with both upsampling and downsampling, we really need some kind of filtering in order to retain just the spectrum of our original signal after the resampling operation. From the previous discussion, it is clear that for the

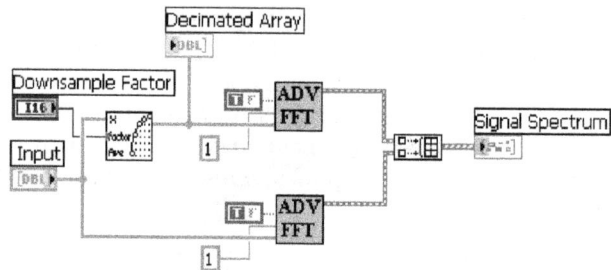

Downsample.vi keeps only 1 out of M samples of input.

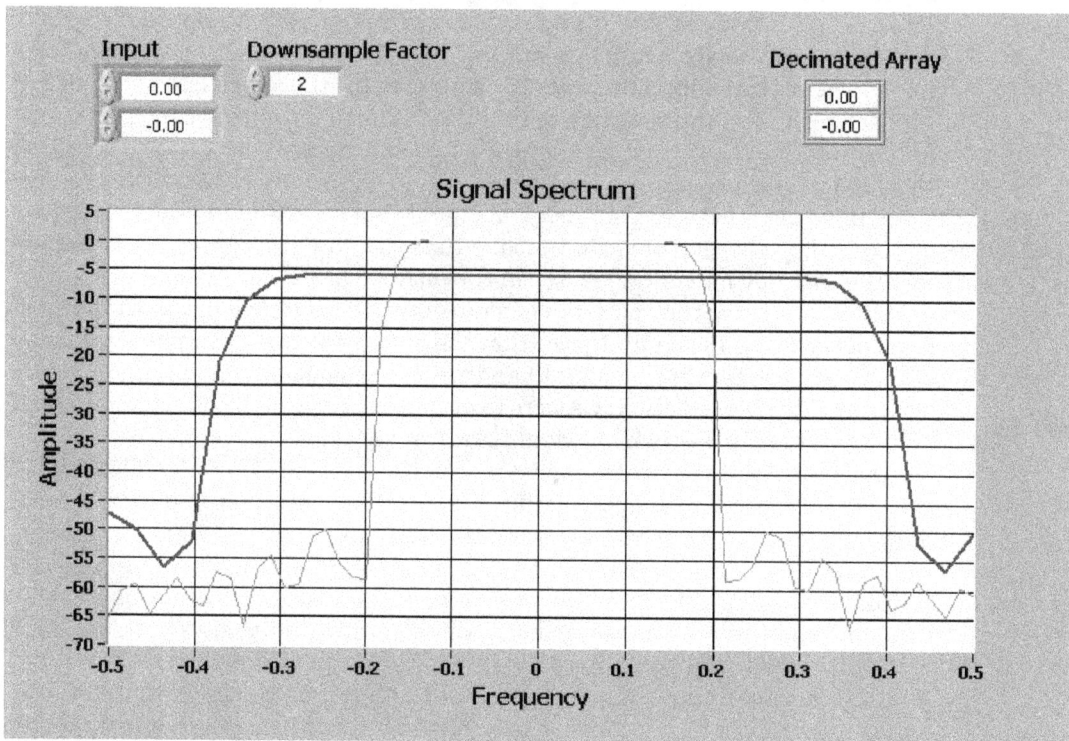

Figure 5.3   Block diagram and front panel for downsample.vi.

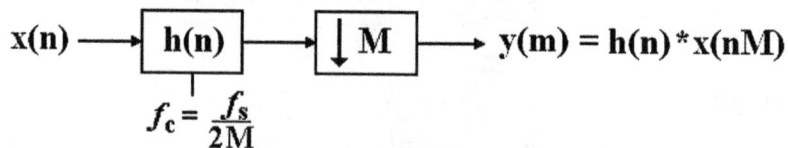

Figure 5.4   Flowchart of downsampling operation.

$$x(n) \longrightarrow \boxed{\uparrow \ \text{I}} \longrightarrow \boxed{h(n)} \longrightarrow y(m)$$

$$f_c = \frac{f_s}{2} = \frac{f_I}{2I}$$

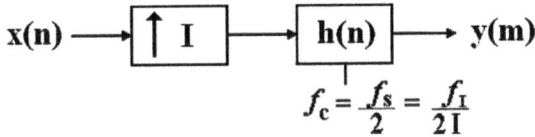

Figure 5.5 Flowchart of upsampling operation.

downsampling operation, we had better filter the signal *first* to make sure we do not get any aliasing. For the upsampling case, there is no aliasing, but if we apply the filter before the upsample operation, the additional spectral replications will still show up unaffected. So for upsampling, the filter should be applied *afterward*.

Figures 5.4 and 5.5 show the flowcharts of the downsampling and upsampling operations. In this traditional type of configuration, the filter in the downsampling step takes place at the input rate, while the filter in the upsampling must take place at the higher output rate. As we saw in Sec. 5.2, the downsampling filter cutoff must limit the input signal bandwidth to $f_s/2M$. The upsampling filter, on the other hand, must have a cutoff equal to half the original sample rate, or half the interpolated sample rate f_I divided by $2I$.

So far the resampling operations we have seen were either up by an integer or down by an integer. For noninteger rates the upsampling and downsampling can be combined to achieve nearly any arbitrary rate. For instance, up by 12, then down by 5 to approximate a rate of 2.35. Of course, a more accurate ratio would be up by 235, down by 100—at the expense of a lot more computations. When combining the two operations, an appropriate filter can be chosen such that a single filter both removes the extra spectral replications from the upsampling *and* prevents aliasing in the downsampling. Since the upsampling requires the filtering afterward and the downsampling requires filtering beforehand, the upsampling should always be performed first. This combined operation is shown in Fig. 5.6.

Combining the upsampling and downsampling as in Fig. 5.6 has obvious computational savings. Performing the two operations separately would require a filter for each operation. The combination allows us to use a single filter to both remove the extra spectral replications from the upsampling and prevent aliasing from the subsequent downsampling. In order for this single filter to remove the correct portion of the spectrum, the cutoff must be chosen as the smaller of $f_I/2M$ or $f_I/2I$. While using a single filter can certainly save computation time, for even greater savings, there are some special conditions involved in the resampled sequence that we can take advantage of. For an explanation of this, let us take a closer look at the upsampling case.

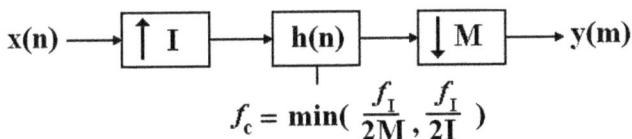

$$x(n) \longrightarrow \boxed{\uparrow \ \text{I}} \longrightarrow \boxed{h(n)} \longrightarrow \boxed{\downarrow \ \text{M}} \longrightarrow y(m)$$

$$f_c = \min\left(\frac{f_I}{2M}, \frac{f_I}{2I} \right)$$

Figure 5.6 Rational ratio resampling.

As we saw in Sec. 5.1, upsampling stuffs $I-1$ zeros between each sample of the input sequence. It is a terrible waste of time to run zeros through the filter when we know exactly where they are. We can eliminate the computations on the zero-valued inputs if we very carefully rearrange the way our input sequence flows through the filter. In fact, we do not have to zero-stuff the input sequence at all and the sequence can still be upsampled by the very nature of the filtering. This type of filter is known as a polyphase filter and is discussed in Sec. 5.3.2. A similar situation exists for the downsampling case where we are first filtering the signal but then throwing away every Mth value. Both cases will be discussed along with a LabVIEW application for performing the polyphase interpolation. There is also another interesting type of resampling filter known as the halfband filter. This filter has the interesting property that every other time domain coefficient is 0, and this can be exploited to cut the required computations in half [2]. Now we will take a look at these two efficient resampling filters.

5.3.1 Halfband filters

As previously mentioned, the halfband filter has the property that every alternating impulse response value is 0 (except the zeroth index sample). The halfband filter is computed with the same filter design techniques that we have already seen (windowing or Parks-McClellan) by setting the filter cutoff frequency to $^1/_4$ of the sample rate.

Figures 5.7 and 5.8 show the block diagram and front panel of Halfband.vi. From the block diagram, we can see that the cutoff frequency is set to $^1/_4$ of the sample rate and the stop frequency is computed from the cutoff and the pass frequency. These values are passed to the MPR.vi and the filter impulse response is shown in Fig. 5.8. The impulse response clearly has the property that every other value is 0. We can take advantage of these zeros to cut the required number of multiplications in half. Obviously the halfband filter implies that the change in sample rate is $2x$ either up or down. For larger changes in the sample rate, stages of these $2x$ resamplers followed by halfband filters can be combined.

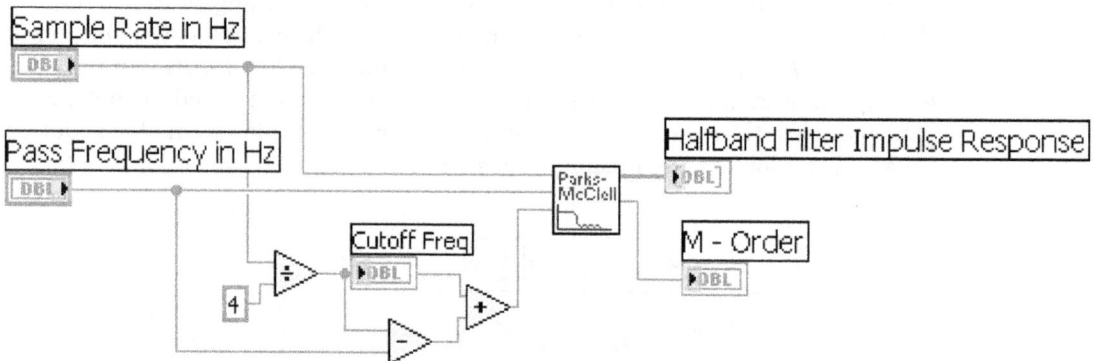

Figure 5.7   Halfband.vi block diagram.

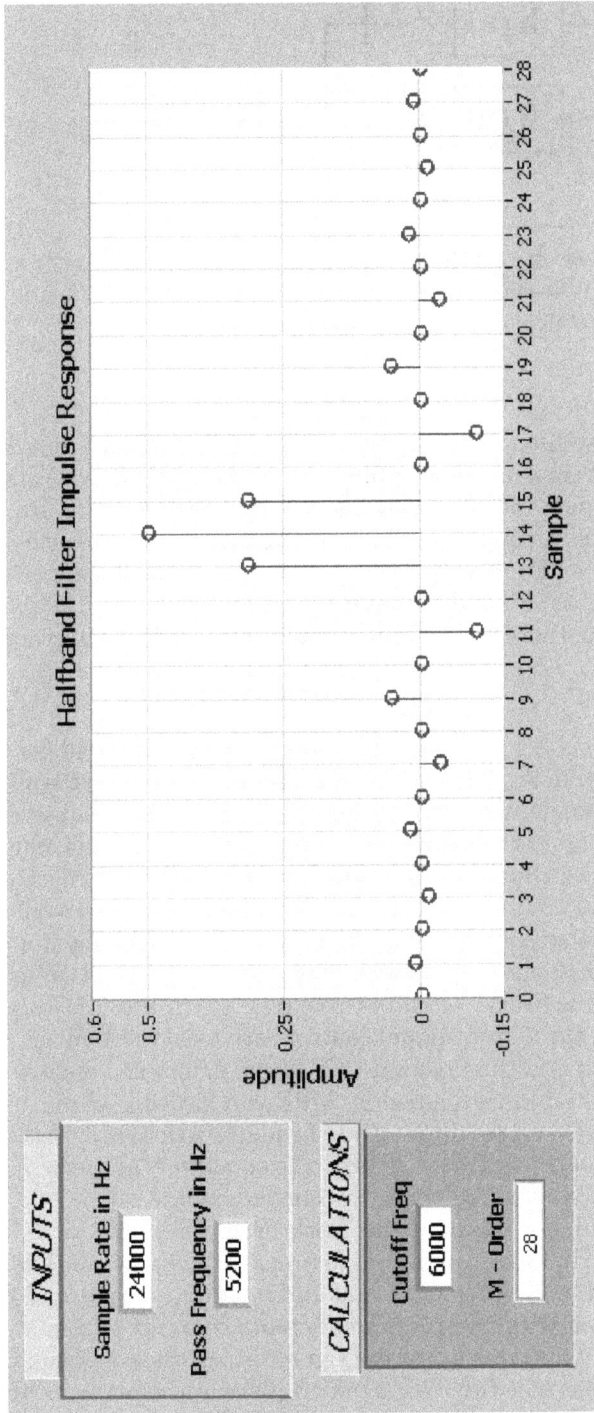

Figure 5.8 Halfband.vi front panel.

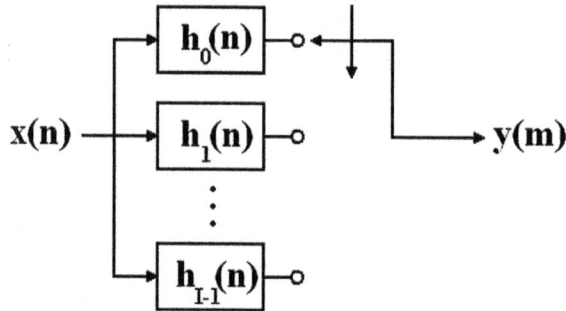

Figure 5.9 Flowchart of a polyphase interpolator.

5.3.2 Polyphase filters

A polyphase filter takes advantage of the fact that computing filter outputs for some inputs in a resampling filter are unnecessary. Specifically, the added zeros in the upsampler and every Mth input in the downsampler are not required for the computation. Using this technique, the polyphase approach can be an extremely efficient filter. First, a filter is designed in the typical manner, next the coefficients are separated into individual filter banks as shown in Fig. 5.9. In a polyphase interpolator, there will be I filter banks with coefficients given by

$$h_k(m) = h[(n + k)I] \qquad (5.1)$$

where $h(n)$ is the original filter before rearranging and $k = 0, \ldots, I - 1$.

You will notice that there is no zero-stuffing involved in the interpolator in Fig. 5.9. The same input $x(n)$ is now being passed to all filter banks and the output commutator chooses the appropriate filter output [3]. In this manner, there are I outputs for each input and any given output requires only $1/I$ times the original number of calculations per output. We will verify this in a moment once we see more about how this polyphase partitioning works by looking at a LabVIEW VI.

Figure 5.10 shows the block diagram of our LabVIEW polyphase interpolator. The VI is broken up into three frames. In the first frame at the top of the figure, the three front-panel controls are used to calculate the necessary filter parameters and then passed to MPR.vi. Afterward we want to check that the length of the filter designed by MPR is a multiple of our interpolation rate so that we can break the filter up into I equal length banks. Notice that the computed cutoff frequency is always going to be equal to $\frac{1}{2}$ of the original sample rate, or, said another way, $1/2I$ times the output sample rate.

In the middle frame of the block diagram in Fig. 5.10, the input signal is upsampled in the conventional manner for comparison sake. That is, $I - 1$ zeros are stuffed in between each real and imaginary input and the resulting signal is convolved with the filter coefficients computed from the Parks-McClellan algorithm. This means that the signal that is passed through the filter is already upsampled, so the filtering is taking place at the higher output rate. If the original input is K samples long and the filter is N samples long, the entire

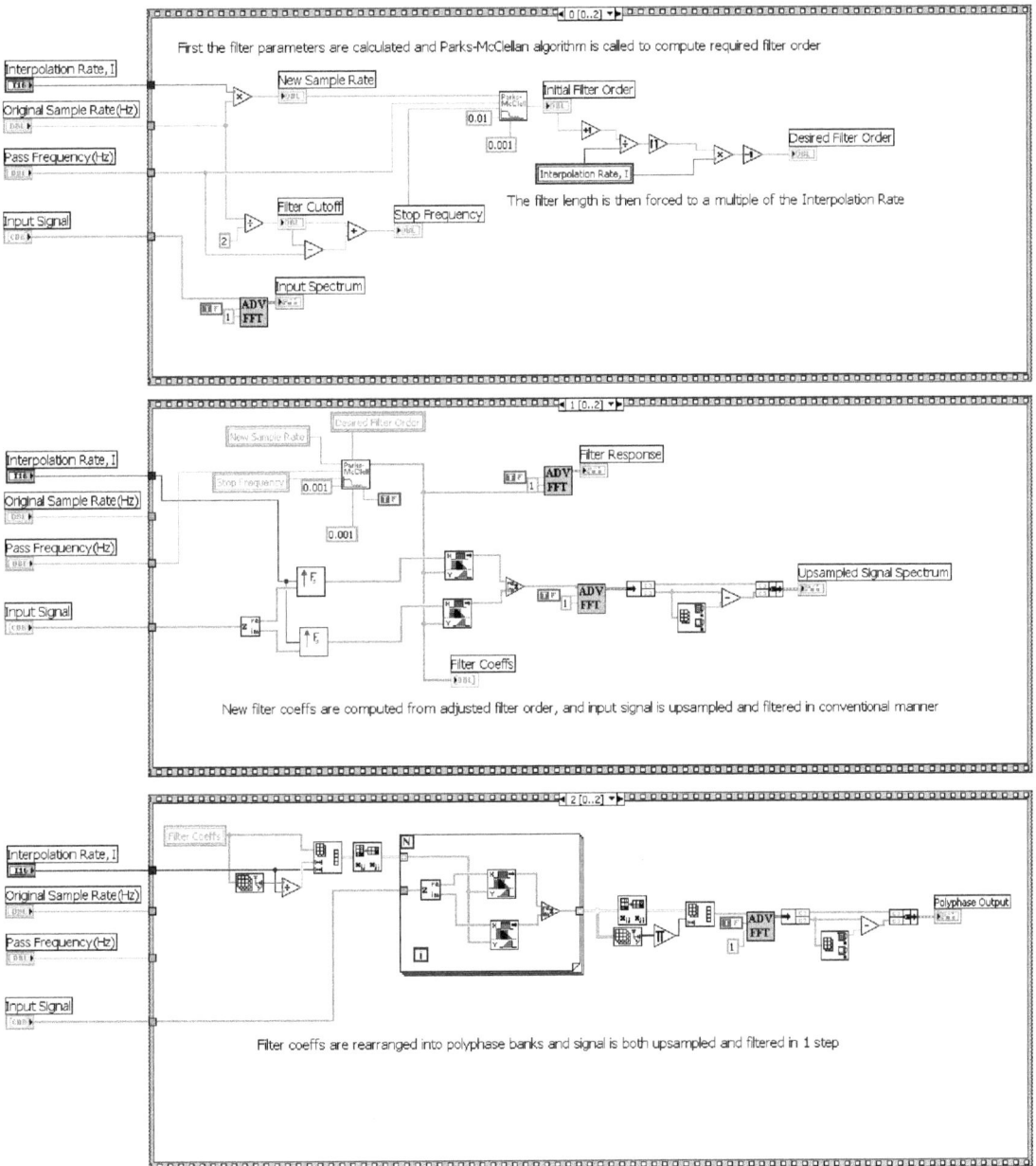

Figure 5.10   PolyphaseInterpolator.vi block diagram.

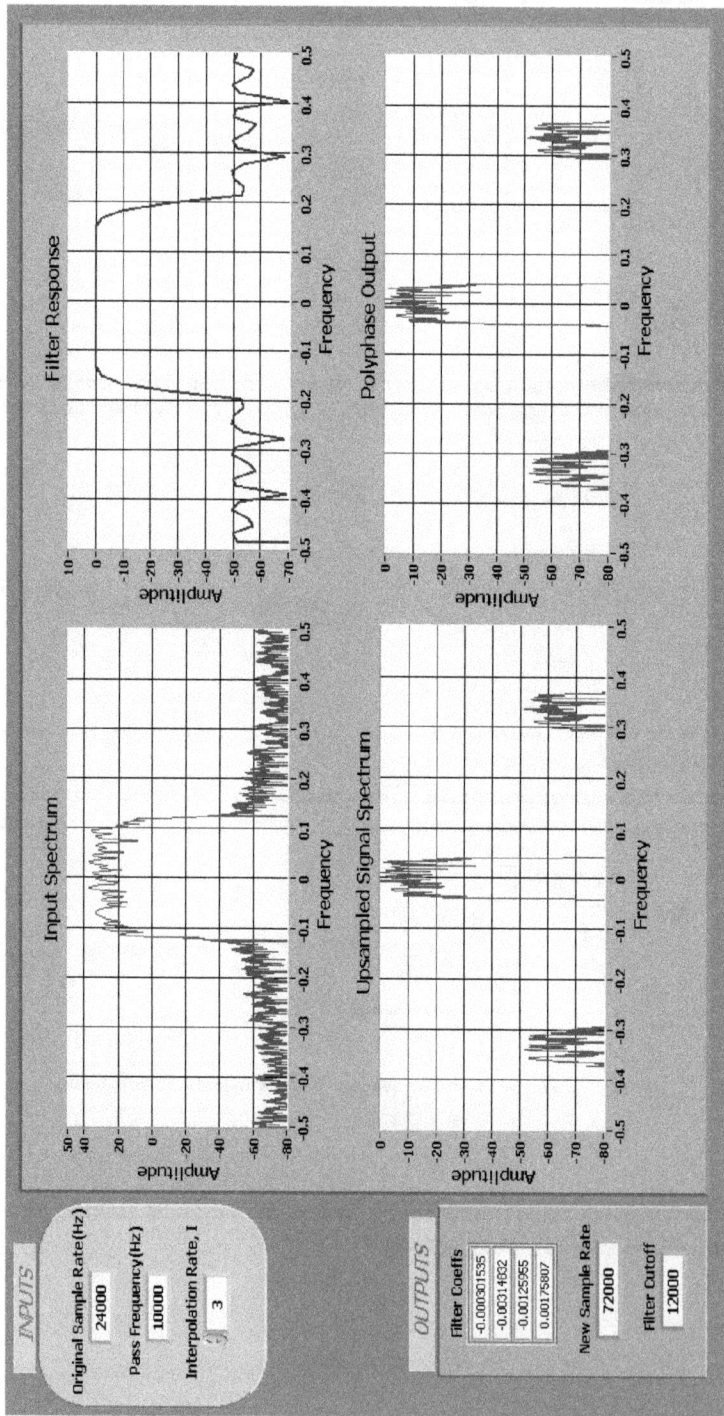

Figure 5.11 PolyphaseInterpolator.vi front panel.

convolution requires $K + N - 1$ multiplications (complex). Now if the input is first upsampled by I, the convolution now requires $I * K + N - 1$ multiplications. On a per output basis, this traditional upsampling followed by filtering requires N (complex) multiplications per output. In a polyphase approach, the filter is split into banks of length N/I and each output thus requires fewer computations.

Finally in the last frame, the polyphase interpolation is performed. The first step is to rearrange the filter coefficients into I filter banks as in Eq. (4.9) by using the reshape and transpose operators. Now the input sequence is convolved with each bank of filter coefficients and the outputs are then interleaved to produce the appropriate filtered sequence.

Figure 5.11 shows us the front panel of our polyphase interpolator. The input spectrum and filter response are shown at the top and the conventional upsampling and polyphase outputs are shown at the bottom. The two output spectra appear to match. This example should convince us that the polyphase partitioning of the filter coefficients does indeed produce the same sequence as the typical upsample or filter *and* the operation is done at a significant savings per output, especially if the interpolation rate is large.

From the preceding discussion, it should also be a short leap to imagine that a polyphase downsampler can be built using the same principles. Namely, the downsampling can occur *before* the filtering and the filter can be broken into M banks. By doing this, the filter will be operating on a sequence with a lower sample rate, thus requiring fewer computations per output. At this point, we will forego actually building a polyphase VI for the downsampler. After studying the structure for the interpolator, it should be a small matter for the reader to transform the interpolator into a polyphase downsampler.

Summary

This chapter has presented some methods for performing sample rate changes. Rate changing can be a very simple operation involving either the insertion of 0 values into the input sequence or the removal of some of the input values. Two efficient rate-changing filters, the halfband filter and polyphase filter, were built in LabVIEW. Only the polyphase interpolator was shown and it was left up to the reader to develop the polyphase downsampler. Of course, there are many more efficient filtering schemes such as Hilbert and Hogenaur filters that are not covered here. The reader is encouraged to read Ref. [3] for further information on efficient resampling filters.

References

1. Oppenheim, A. V., R. W. Schafer, and J. R. Buck, *Discrete-Time Signal Processing,* 2d ed., Prentice-Hall, Upper Saddle River, NJ, 1998.
2. Lyons, R. G., *Understanding Digital Signal Processing,* Prentice-Hall, Upper Saddle River, NJ, 2001.
3. Harris, F., *Multirate Digital Signal Processing,* Prentice-Hall, Upper Saddle River, NJ, 2004.

Generating Signals
with LabVIEW

In this chapter we see how to generate some common signals that are encountered in various communication systems. This chapter presents some of the more basic signals that will be useful for our work in digital signal processing (DSP) and digital communications. Luckily LabVIEW already incorporates some built-in functions for generating most of these simple signals. In DSP and communications, we have many uses for sinusoids, and we will also make use of the impulse, ramp, and white gaussian noise functions. Figure 6.1 shows the location in the functions palette where the various signal generation virtual instruments (VIs) are located.

6.1 Basic Functions

Some of the most useful signals in DSP also happen to be some of the simplest. Three of the most widely used signals are the impulse, ramp, and step. Figure 6.2 shows the block diagram of BasicFcns.vi, which uses the built-in LabVIEW functions for generating these three signals. The actual VIs are named Impulse Pattern.vi, Ramp Pattern.vi, and Pulse Pattern.vi.

The impulse function has the equation given in Eq. (6.1) and can be used as the input signal to extract the filter coefficients from a filtering routine. It can also be used to generate a flat noise spectrum. This VI has three inputs: the number of samples, the impulse amplitude, and the delay.

$$\delta[n - n_D] = \begin{cases} 1 & n = n_D \\ 0 & n \neq n_D \end{cases} \tag{6.1}$$

The ramp function will be extremely useful in generating any linear type of sequence. It was used previously in this book to generate the frequency axis for displaying the spectrum of signals we have examined. This VI also has three

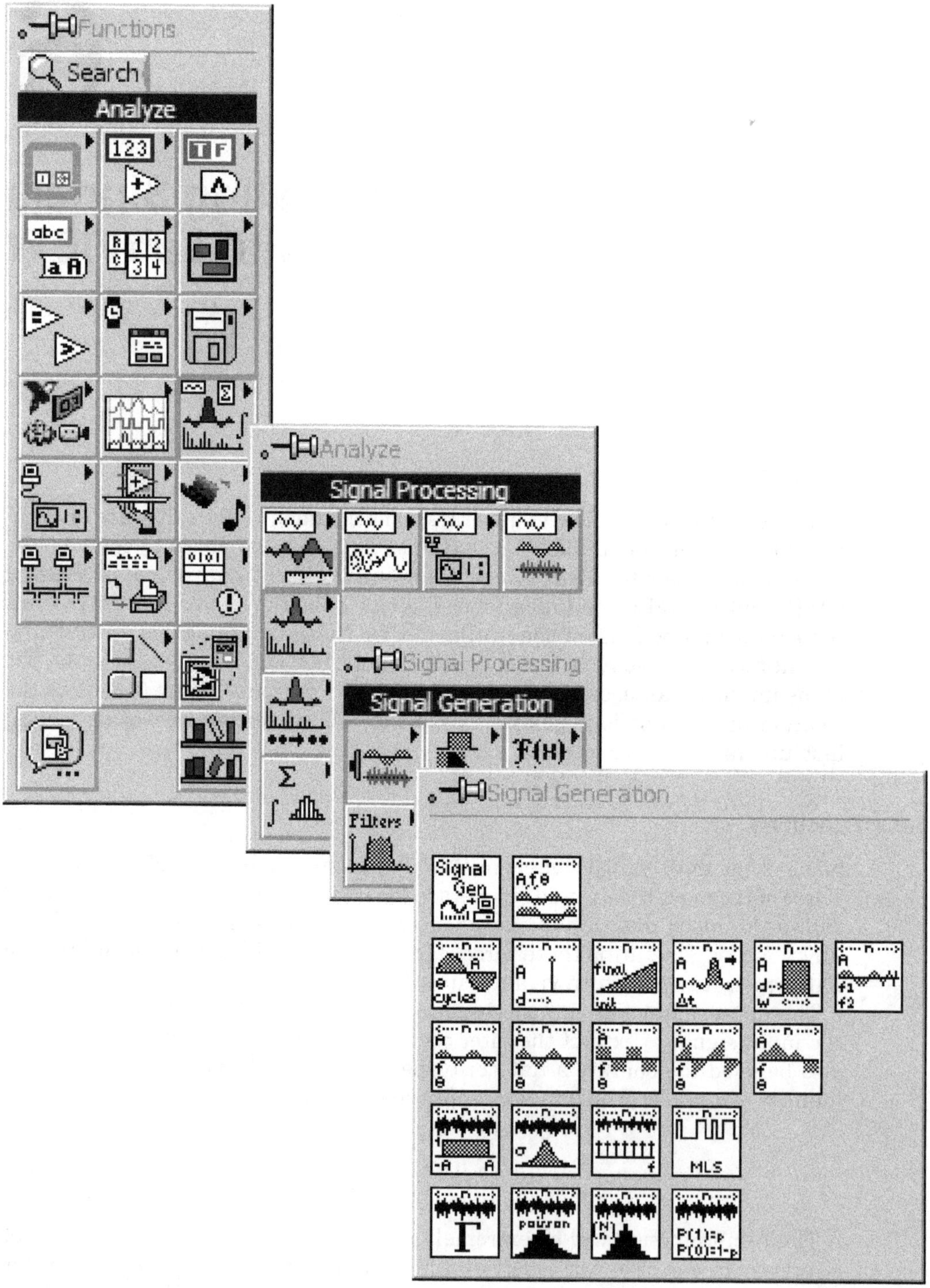

Figure 6.1 Signal generation VIs.

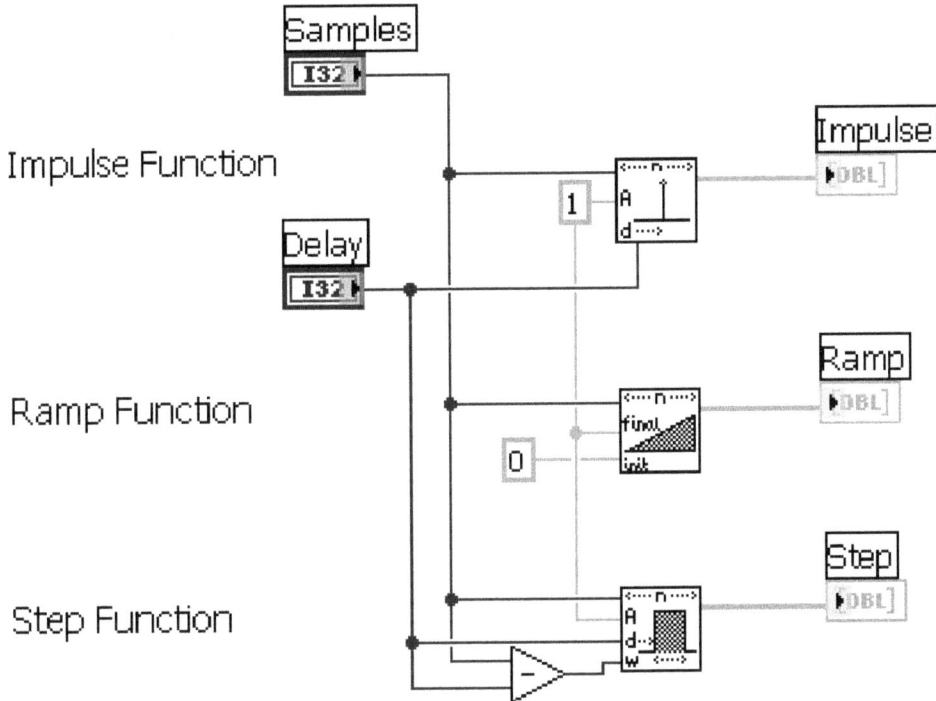

Figure 6.2 BasicFcns.vi block diagram.

inputs: the number of samples, the start value, and the stop value. An interesting property of Ramp Pattern.vi is that it is capable of generating either positive or negative sloped ramps depending on the relationship between the start and stop values.

The step function or Pulse Pattern.vi has four inputs: the number of samples, amplitude, delay, and width. The equation for a step function is shown in Eq. (6.2) and the time-domain plots of all three functions are shown in Fig. 6.3.

$$u[n - n_D] = \begin{cases} 1 & n = n_D \\ 0 & n \neq n_D \end{cases} \tag{6.2}$$

6.2 Sinusoids

Another very important class of signals for DSP and digital communications is the sinusoids. We have already seen some examples of sinusoids such as the raised cosine and a brief glimpse of the sinc function. Here we will also look at generating a complex mixer and an example of frequency modulation in the form of a chirp sequence.

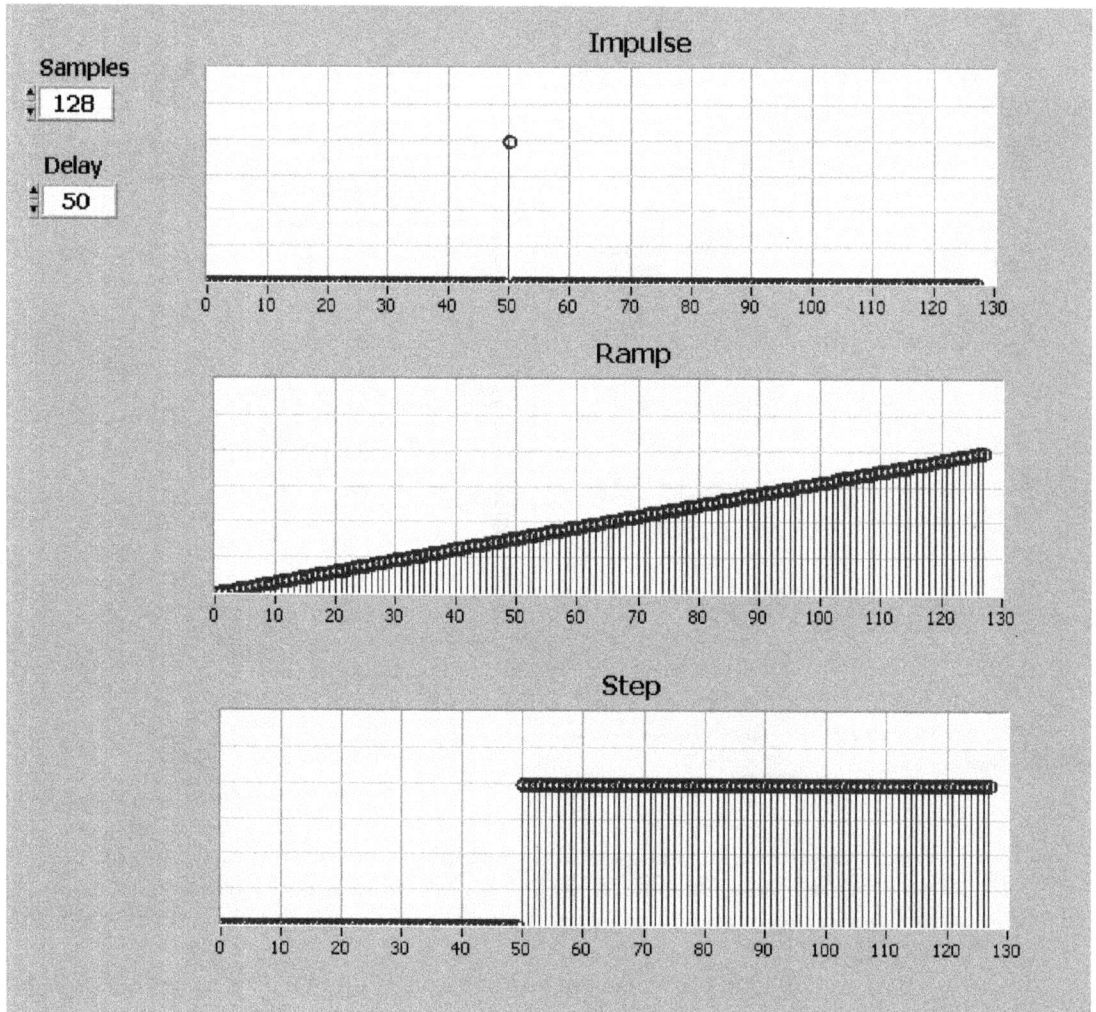

Figure 6.3 BasicFcns.vi front panel.

6.2.1 Complex mixer

Since LabVIEW includes a routine to generate a sine pattern, we will use it as the basis for building our complex mixer. We will also need to know the data sample rate, desired length of the mixer, and the mixer frequency. The LabVIEW sine pattern generates a sine wave of a given number of cycles (fractional cycles allowed) in a specified number of samples. The block diagram shown in Fig. 6.4 divides the mixer length by the data sample rate to give the time length in seconds. The time is then multiplied by the mixer frequency to give the number of cycles. We can simultaneously generate a cosine wave by setting the phase offset of the sine pattern generator to 90°. Finally for compactness, we can combine

Generates a complex sinusoidal mixer given the desired length, frequency and sample rate :

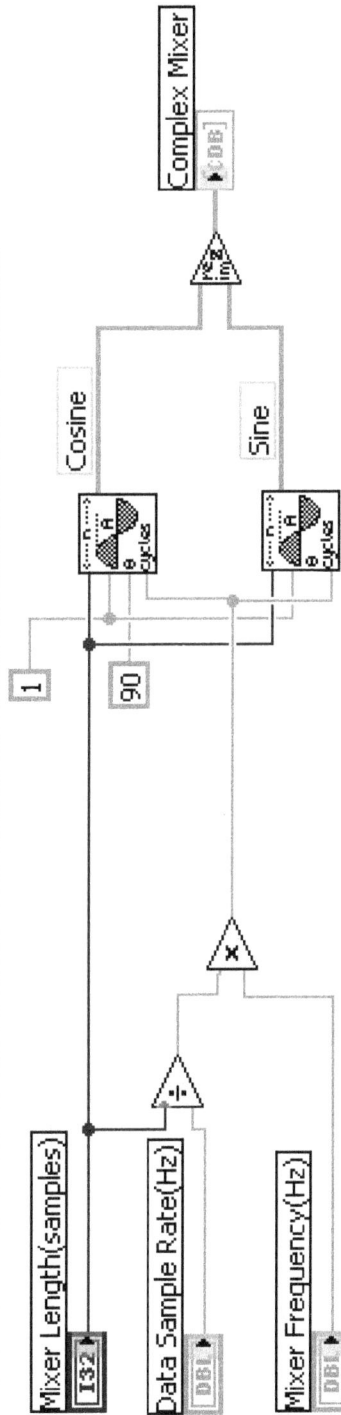

Figure 6.4 Complex mixer generation.

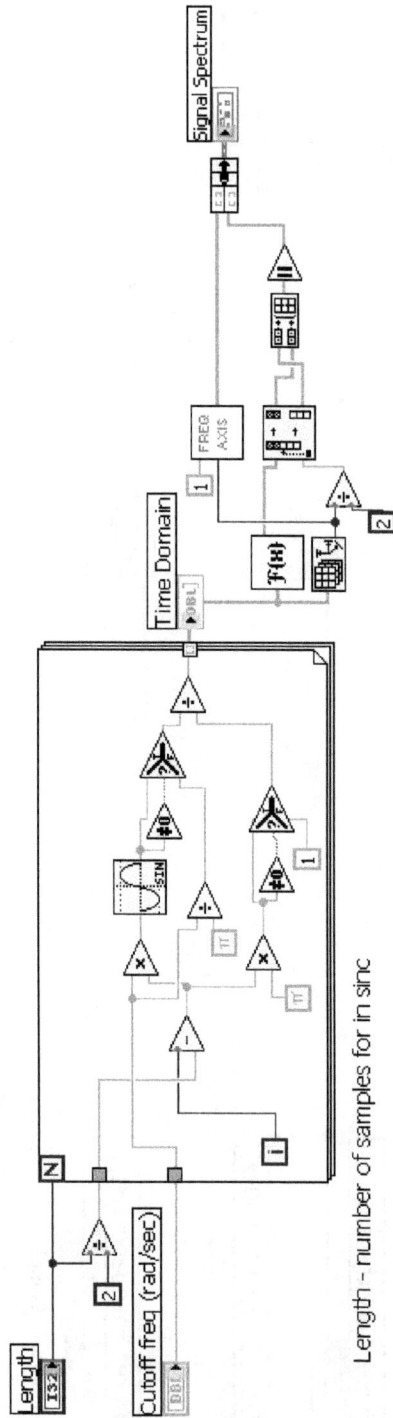

Length – number of samples for in sinc

Cutoff freq (rad/sec) – cutoff = (pass freq + stop freq) / 2

Figure 6.5   SincFcn.vi block diagram.

the sine and cosine into real and imaginary parts of a complex number, thereby creating our complex mixer.

6.2.2 Sinc function

Another useful function that was briefly introduced in Chap. 4 is the sinc function VI called SincFcn.vi. This VI computes the sinc waveform from the formula [1]:

$$h[n] = \frac{\sin[\omega_C(n - M/2)]}{\pi(n - M/2)} \tag{6.3}$$

The block diagram that computes Eq. (6.3) is shown in Fig. 6.5. You might notice that there is a singular point where the sinc function would not be well behaved if we did not control it somehow. That point is where $n = M/2$, which results in 0/0 division. The VI handles this special case with two case selectors and sets the sinc amplitude to a specific value at the discontinuity.

The front panel of SincFcn.vi is shown in Fig. 6.6. We used this truncated sinc (remember that the real sinc function has infinite duration) back in Chap. 4

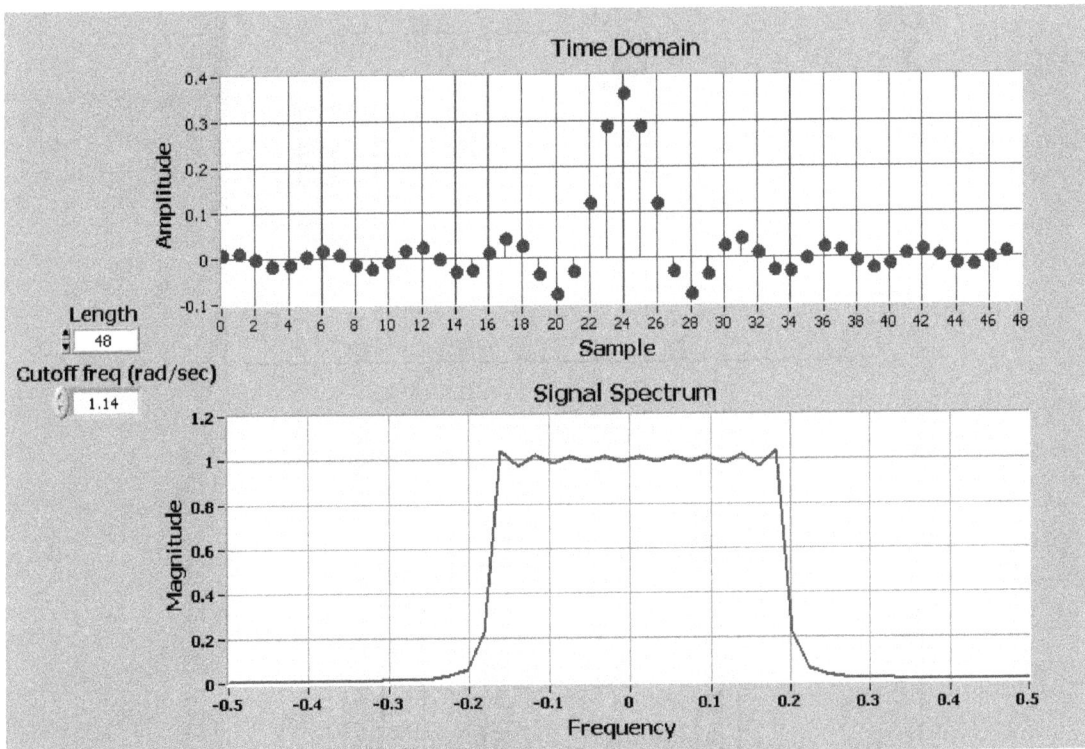

Figure 6.6   SincFcn.vi front panel.

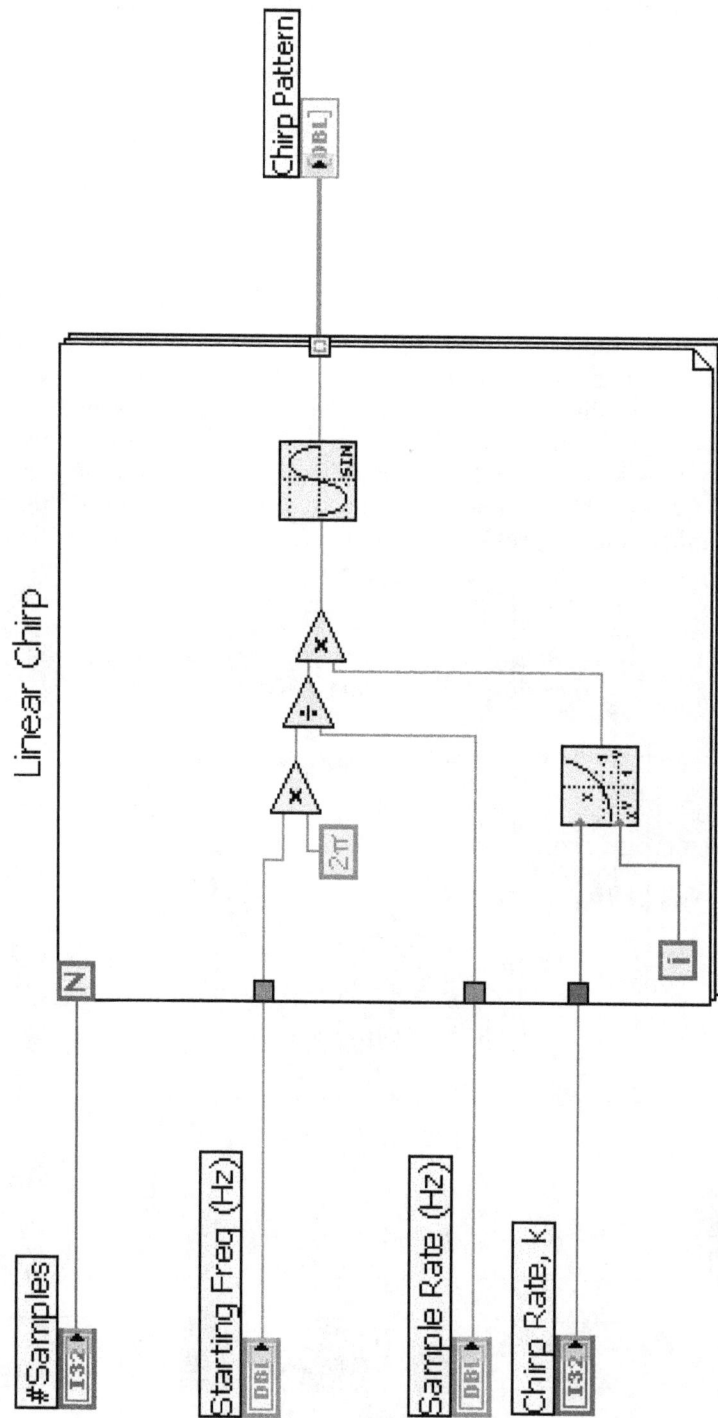

Figure 6.7 Chirp.vi block diagram.

when we needed a prototype for filter design by windowing. As you increase the length of the sinc waveform you will see that the spectrum gets closer and closer to the ideal low-pass filter.

6.2.3 Chirp sequence

The chirp sequence is an interesting signal that has applications in sonar, radar, and spread spectrum communications. There are two different forms of chirp sequence, linear and exponential, which refer to the rate at which the chirp signal frequency changes. The simpler of the two is the linear chirp, which we will examine here. Equation (6.4) gives the formula for a linear chirp sequence and the block diagram in Fig. 6.7 shows how this can be implemented.

$$w[n] = \sin\left(\frac{2\pi f}{f_S} n^k \right) \tag{6.4}$$

Chirp.vi linearly increases the frequency of the waveform beginning at the start frequency and continuing for the number of samples. The instantaneous frequency of the waveform is given by the derivative of the phase with respect to time. The generated chirp pattern is shown in Fig. 6.8.

6.3 Generating Channel Models

An important part of digital communications is simulating signal recovery in the face of noise and other channel impairments. In order to really put our communication system through its paces we need to be able to generate a model of the channel. The amplitude and phase variations imposed by the channel on the received signal can generally be termed fading. There are several different types of fading as well as several different models to simulate those fading conditions. For simplicity, we will consider only small-scale flat fading. Reference [2] tells us that for 2-D isotropic scattering environments, the magnitude of the received complex envelope $g(t)$ can be modeled as a Rayleigh random variable at any given time and the phase is uniformly distributed from $-\pi$ to π. Section 6.3.1 explains how to build a Rayleigh model of our channel. In addition to fading, there are other sources of noise to consider. For example, the receiver will have some thermal noise and there may be interference from other radio frequency (RF) signals. We will model these extraneous noise sources in Sec. 6.3.2 as gaussian noise.

6.3.1 Rayleigh fading

So how do we generate a Rayleigh envelope in LabVIEW? Figure 6.1 showed that LabVIEW contains a function for generating white gaussian noise (as well as uniform white noise, Gamma, Poisson, and binomial noise). Well, again Refs. [2, 3] tell us that if we have a random variable R defined as

$$R = \sqrt{[g_I(t)]^2 + [g_Q(t)]^2} \tag{6.5}$$

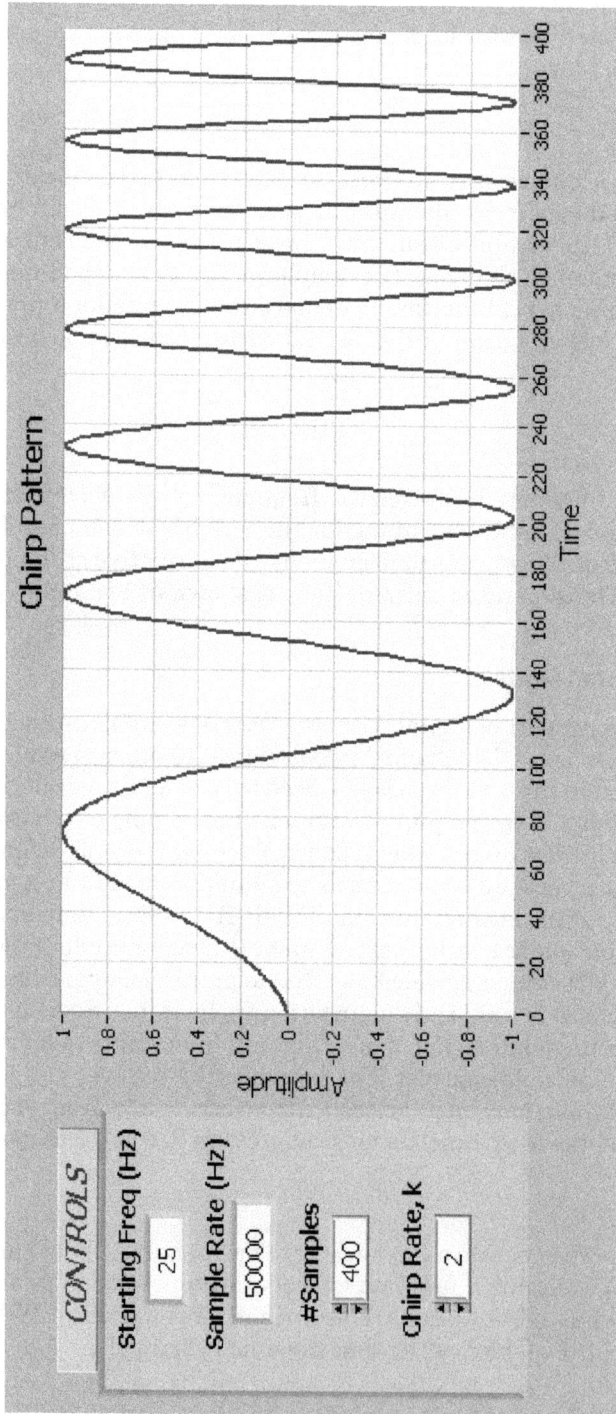

Figure 6.8 Chirp.vi front panel.

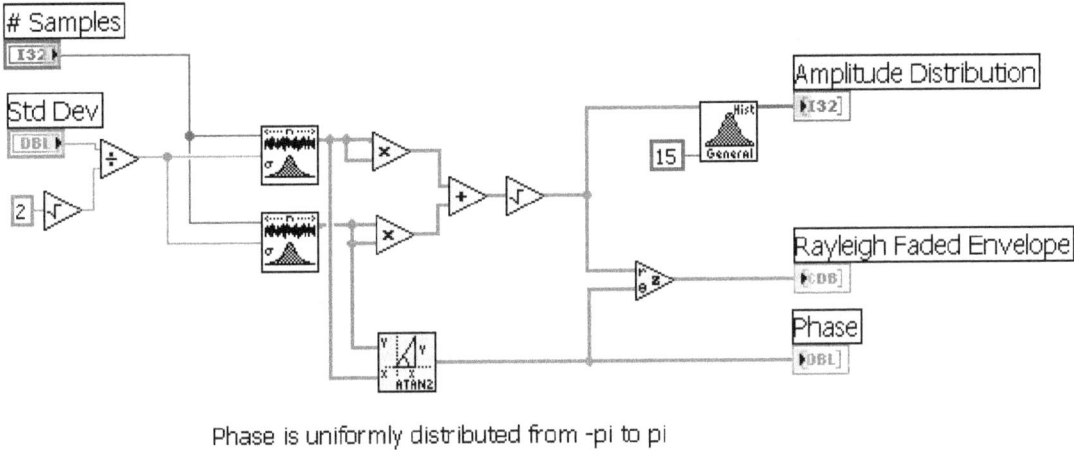

Phase is uniformly distributed from -pi to pi

Figure 6.9   Rayleigh.vi block diagram.

then R will be Rayleigh distributed as long as $g_I(t)$ and $g_Q(t)$ have zero mean gaussian distributions (and the number of scatter paths is large). Relating Eq. (6.5) to our recovered signal, $g(t) = g_I(t) + g_Q(t)$ is the received complex envelope and $R = |g(t)|$ is the magnitude of that envelope. We can now use this information and the white gaussian noise VI to build the block diagram shown in Fig. 6.9.

Rayleigh.vi takes two inputs: the number of samples and the standard deviation of the scattering envelope. Since the standard deviation is the square root of the signal power, we divide by $\sqrt{2}$. That way the real and imaginary parts of the Rayleigh envelope each contain half of the total signal power. The term "scattering" comes from the fact that the small-scale faded envelope is a result of multiple receive paths arising from the transmitted signal being reflected off obstacles. The scattering can be caused by any objects in the environment such as buildings, trees, automobiles, and so on. In Fig. 6.9, two gaussian distributed sequences $g_I(t)$ and $g_Q(t)$ are formed and they are squared, summed, and then the square root yields the Rayleigh envelope. Finally, the arctangent of $g_Q(t)/g_I(t)$ gives us a uniformly distributed phase over the interval $-\pi$ to π. In Fig. 6.10, you can see that the histogram of the complex envelope is indeed Rayleigh distributed.

Now that we have a model for the channel gain magnitude and phase from Rayleigh.vi, we can start to build a fader that will simulate the channel conditions for our modulated signal. To do this, we need to simply generate the desired number of Rayleigh faded paths, sum up their contributions, and multiply the sum by our modulated signal. Since this chapter is focusing on generating the underlying signals, we will wait until Chap. 7 to apply the Rayleigh fading to our digitally modulated signal.

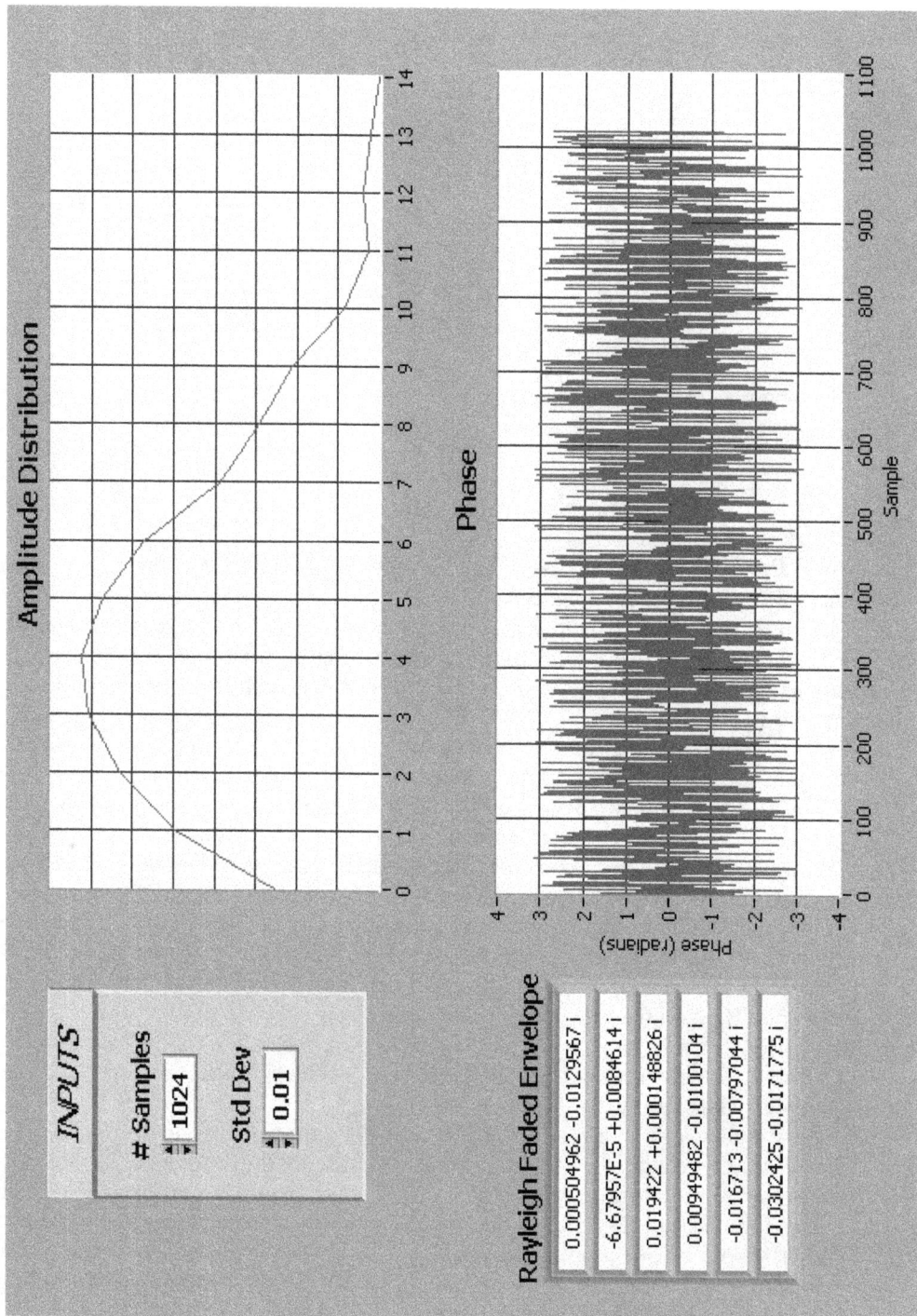

Figure 6.10 Rayleigh.vi front panel.

Keep in mind that the Rayleigh faded envelope does not include the presence of a direct line-of-sight component. This type of distribution has a ricean pdf, of which the Rayleigh distribution is a special case. This section attempts to only show the potential for channel modeling in LabVIEW, the inclusion of a line-of-sight component is left up to the reader.

6.3.2 White gaussian noise

Besides the fading from the channel, there will also be noise from nearby RF signals as well as thermal noise in the receiver itself. We can model these noise sources as white gaussian noise processes, which LabVIEW can easily generate using the same white gaussian noise VI that we saw in Sec. 6.3.1. The block diagram of AWGN.vi is shown in Fig. 6.11 and the front panel is shown in Fig. 6.12.

This VI simply forms either a real or complex white gaussian noise sequence with a given variance. An examination of the noise spectrum shows that the frequency content of the noise is flat, as we would expect for white noise. We will use this VI in later chapters to degrade our transmitted waveform and simulate a noisy environment.

6.4 Generating Symbols

Our last step before simulating the whole digital communication structure will be to generate a digitally modulated signal. That means that we have to choose a digital modulation scheme like QAM, QPSK, PAM, or any number of other modulations. We will also need some bits to actually map to the symbols in our chosen modulation. Producing the bits is as simple as using the LabVIEW random number generator as shown in Fig. 6.13. The random number generator produces a uniform distribution of numbers between 0 and 1. The double precision random numbers are then rounded to the nearest integer and cast to 16-bit integers.

Once we have the bits, we are ready to map them to our modulation symbols. For this discussion, 16-QAM was the chosen modulation scheme. To map the bits to a symbol, a symbol table is formed with the symbols in Gray encoded order and the appropriate symbol is chosen by simply indexing this array of symbols with the bits to transmit [4]. The Gray encoding gives us a reduced risk of a bit error by mapping neighboring symbols to bit values that differ by only a single bit. This means that if a neighboring symbol is chosen by mistake, the error will only be 1 bit. Since the Gray encoding is done in the design stage, this is a very nice trick for reducing error rates with absolutely no cost in terms of computation. Figure 6.14 shows a 16-QAM symbol mapper with a Gray encoded symbol table.

The 16-QAM modulation maps 4 bits to each symbol, and you can see in the figure that the input bit array is split at index 4 with the top 4 bits being converted

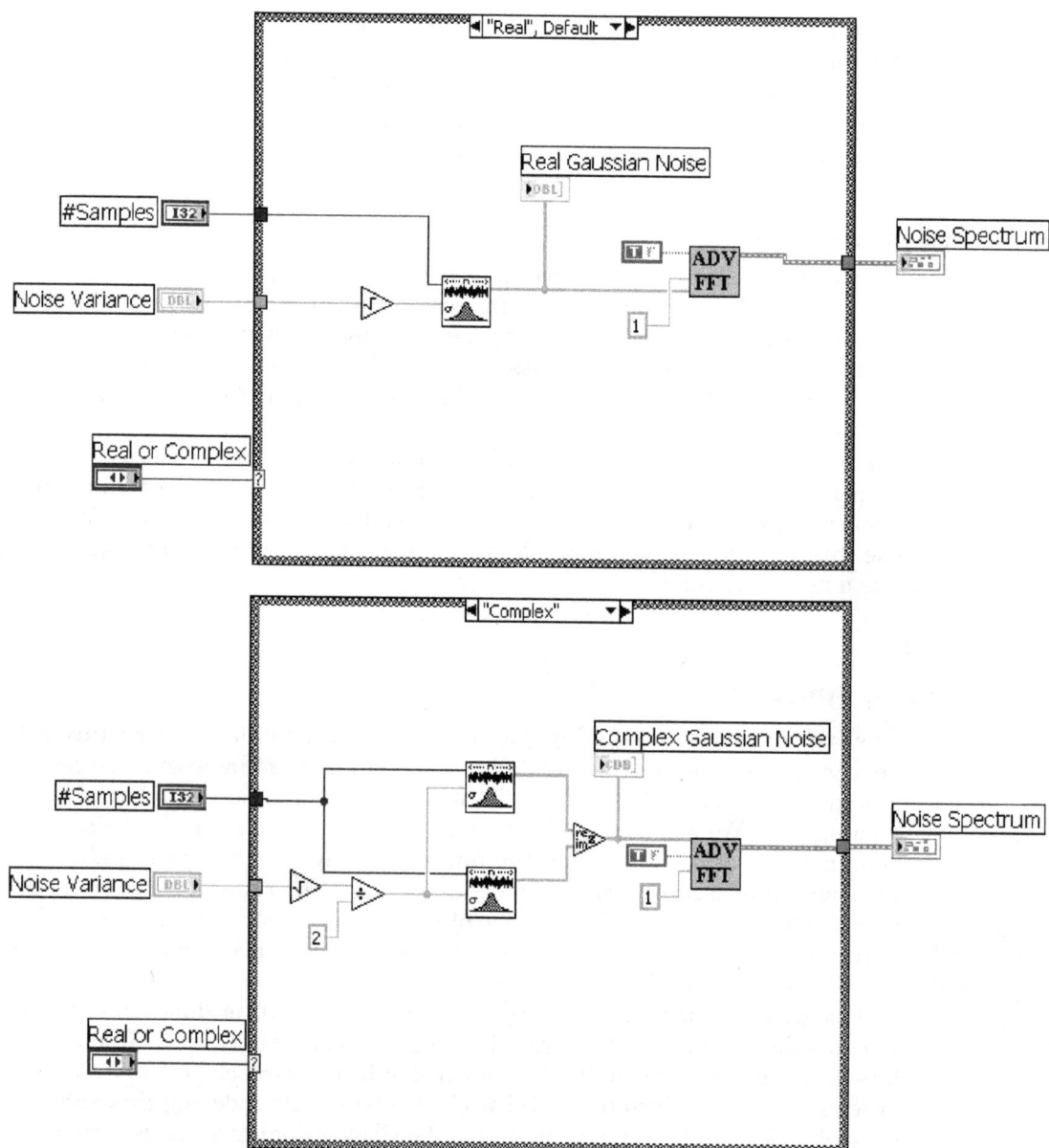

Figure 6.11 AWGN.vi block diagram.

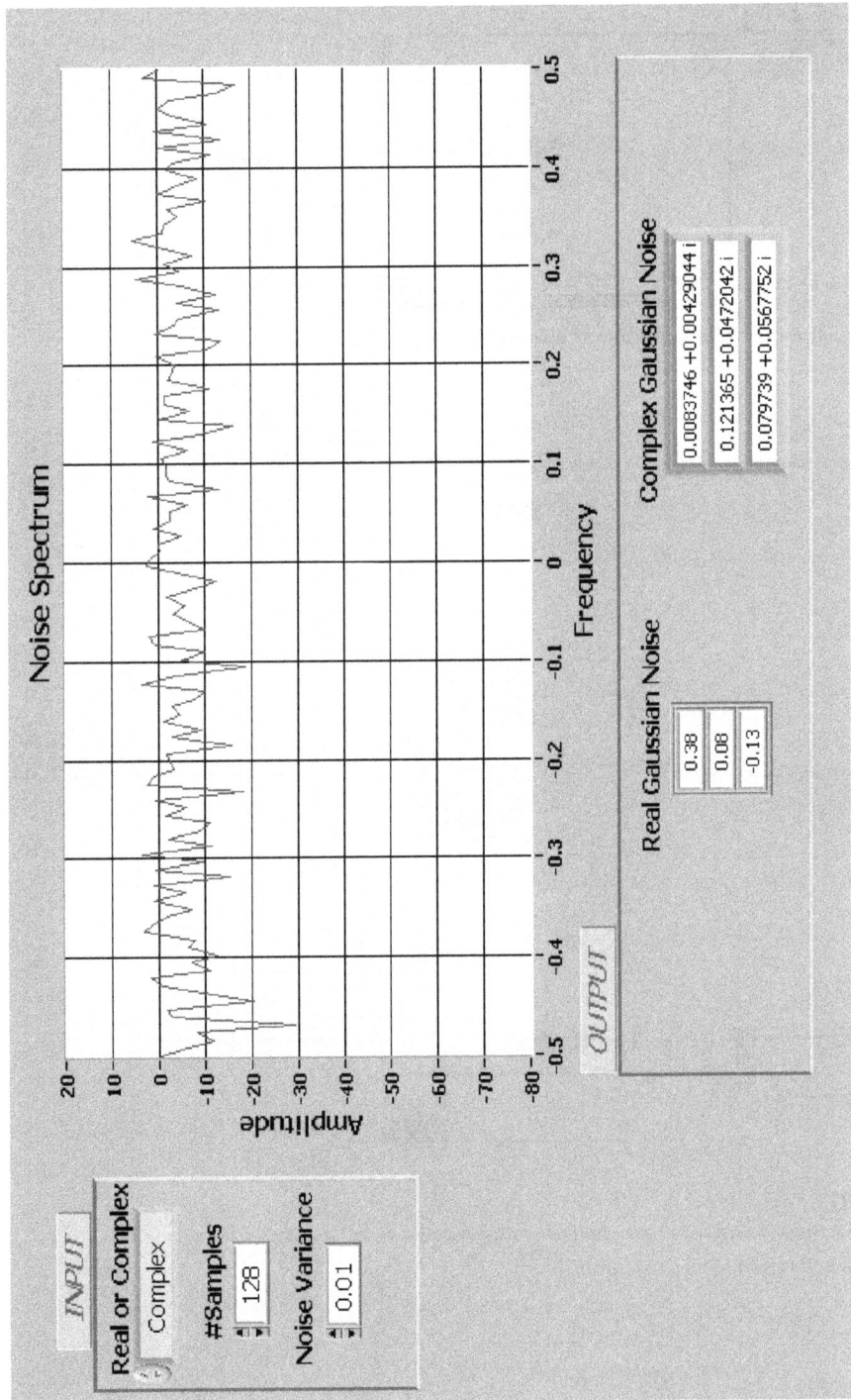

Figure 6.12 AWGN.vi front panel.

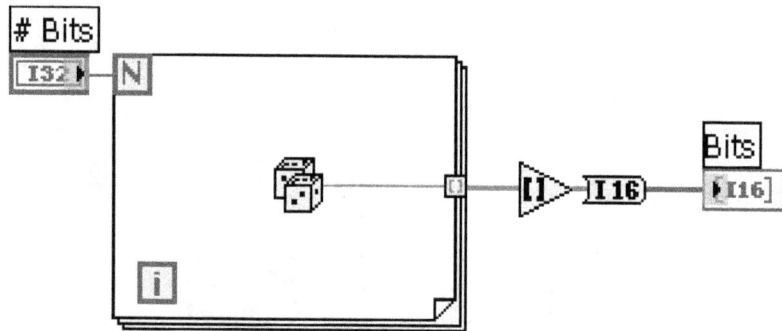

Figure 6.13 GenerateBits.vi block diagram.

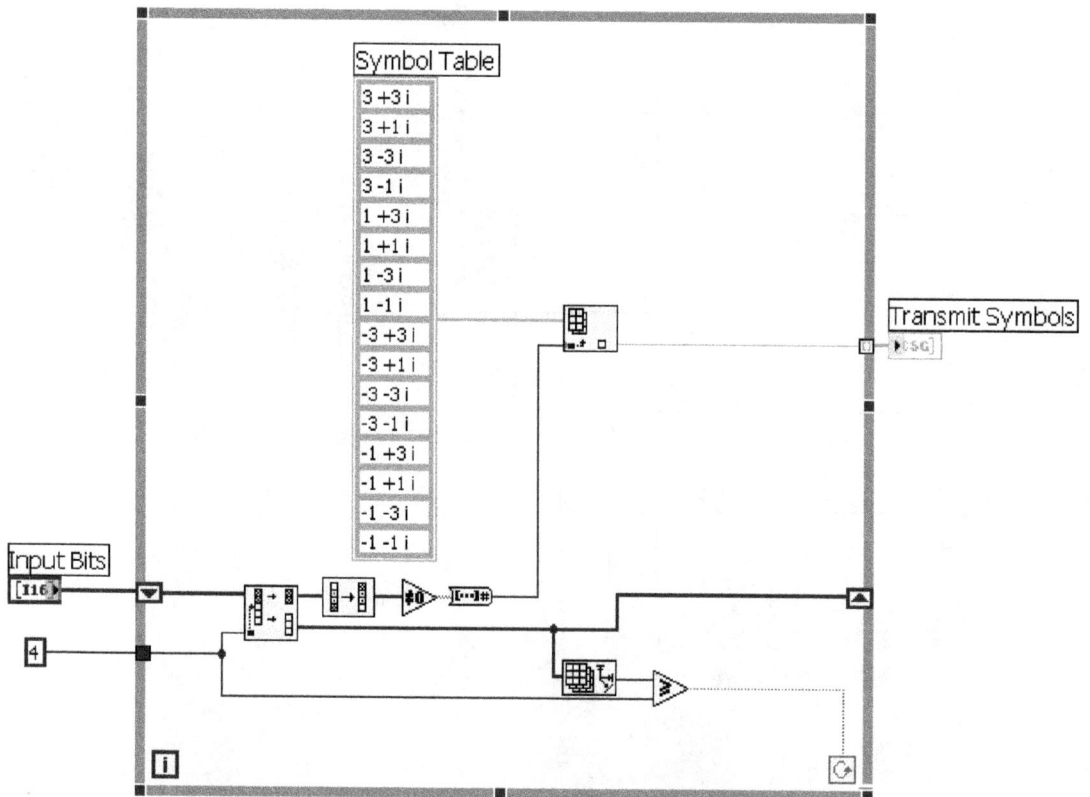

Figure 6.14 SymbolMapper.vi block diagram.

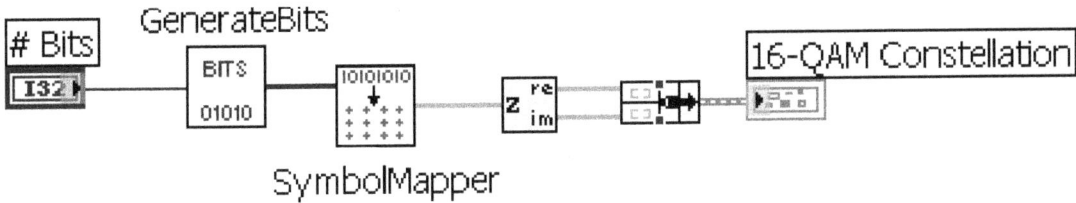

Figure 6.15 Plotting the constellation.

to a numeric index for the symbol table and the remaining bits are passed to a shift register for the next loop iteration. The 4-bit array is first reversed because of the way the boolean array to number function works. We can then view the constellation by bundling the real and imaginary parts of the symbols and creating an X–Y graph as shown in the block diagram in Fig. 6.15. The front panel of the X–Y graph is shown in Fig. 6.16.

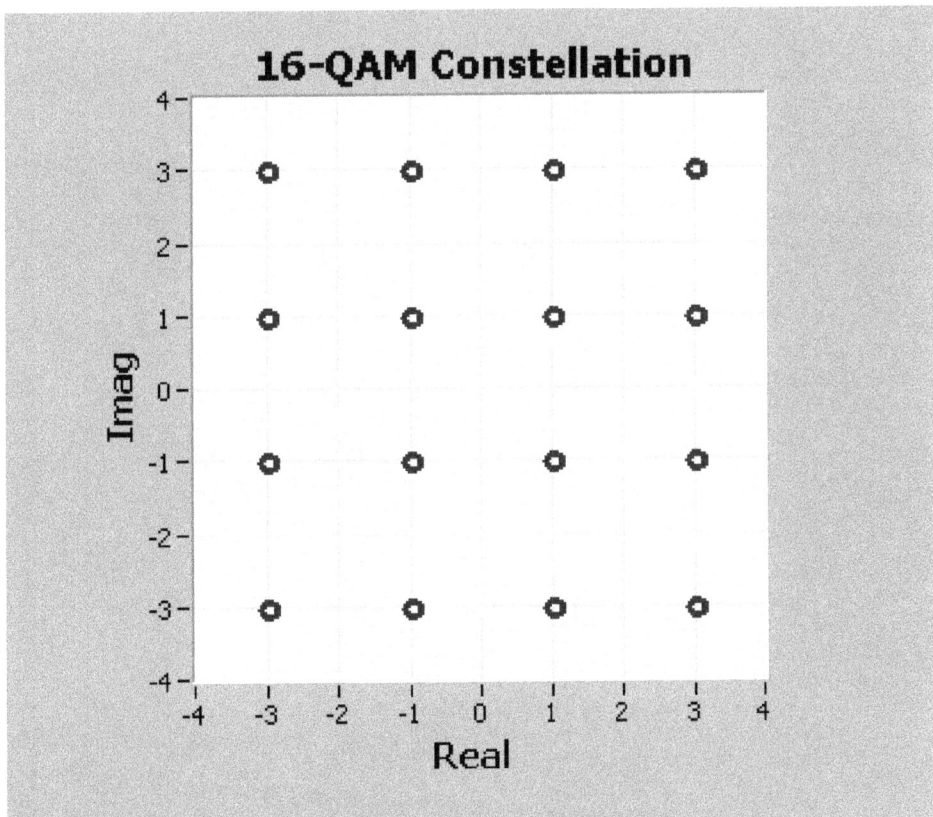

Figure 6.16 16-QAM constellation X–Y graph.

Summary

This chapter has introduced some of the signal generation capabilities of LabVIEW. We started with some relatively basic signal types, then moved from sinusoids to random noise, and finally we saw how to map a bitstream to a 16-QAM constellation. Of course, all of these sections only touched the surface of the LabVIEW or digital communication world, but hopefully there is enough information here to get you started down the path of understanding. In the next few chapters, we will see all of these topics take the shape of a real communication system from bits to bits on both ends.

References

1. Oppenheim, A. V., R. W. Schafer, and J. R. Buck, *Discrete-Time Signal Processing,* 2d ed., Prentice-Hall, Upper Saddle River, NJ, 1998.
2. Stuber, G. L., *Principles of Mobile Communication,* 2d ed., Kluwer Academic Publishers, Boston, MA, 2001.
3. Prabhu, G. S., and P. M. Shankar, "Simulation of Flat Fading Using MATLAB for Classroom Instruction," IEEE Trans. on Education, vol. 45, no. 2, February 2002, pp. 19–25.
4. Proakis, J. G., *Digital Communications,* 4th ed., McGraw-Hill, Boston, MA, 2001.

3

Building a Communication System

Assembling the Pieces

Chapters 1 through 6 have explained most of the basic LabVIEW DSP tools that we will need for dealing with digital communication signals. In this chapter we finally start seeing all those ideas come together into the complete digital communication package. First we will start by building up the transmit waveform as shown in the flowchart in Fig. 7.1. We have already looked at all the virtual instruments (VIs) that we need to form a digitally modulated signal and from the discussions in Chaps. 1 to 6, we should have at least a cursory understanding of the important pieces of the modulator block diagram shown in Fig. 7.1. Now we are ready to assemble those pieces into a working modulator.

The reverse process of the digital modulator is shown in Fig. 7.2. For the most part, the pieces of the demodulator are the same as the modulator; however, some of the demodulation tools have not yet been discussed and will be covered later in this chapter.

Once we have a working modulation/demodulation system, we will simulate a noisy radio frequency (RF) fading channel by building Channel.vi using the channel model from Chap. 6. We will begin with the modulator in Sec. 7.1.

7.1 Modulator

Starting on the left of Fig. 7.1, we already saw (in Chap. 6) a simple VI called GenerateBits.vi to give us a random sequence of bits. Then in Sec. 6.4, we examined a method for mapping those bits to Gray coded points on the 16-QAM constellation. Next we upsample the baseband signal to our desired sample rate and then run the data through the pulse shape filter. Remember that our signal will be prone to intersymbol interference (ISI) if we do not do any kind of pulse shaping. For the shaping filter we have to return to Sec. 4.5 and use NyquistPulse.vi to design either a raised cosine or root-raised cosine window.

bits
1010101010
→ **Symbol Mapping** → **Upsample** → **PSF** → **TX Waveform**

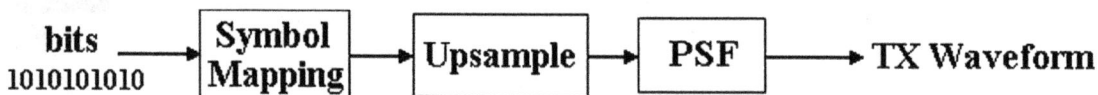

Figure 7.1 Flowchart showing the generation of a digitally modulated waveform.

Now that we have refreshed ourselves on the components of the modulator, let us examine the block diagram of Modulator.vi in Fig. 7.3. On the left are the familiar functions from Chap. 6 for generating bits and mapping them to the 16-QAM constellation. NyquistPulse.vi is then called to design a pulse-shaping filter for a given length, sample rate, symbol rate and roll-off factor (alpha). Here the Nyquist shaping pulse is produced at the desired output sample rate and *not* at the input symbol rate. From Sec. 4.4.2, we know that the delay of this pulse shape filter is $M/2$, where M is the filter order. And we also know that $M = N - 1$, where N is the filter length, thus giving us the filter delay in terms of the length [1]. Now we zero-stuff the real and imaginary parts of the input symbol waveform and convolve the upsampled waveform with the pulse-shaping filter coefficients. The final step is to generate the frequency spectrum with AdvFFT.vi and to shift the y-axis maximum value to 0 dB. Note also that the upsample or filter operation could just as easily be done with the polyphase approach from Chap. 5.

Figure 7.4 shows the front panel of Modulator.vi. Let us look at a couple of interesting points in this figure. First of all, notice that there are five samples from the peak value of the pulse-shape filter out to the first zero crossing on either side. This is a quick and easy way to recognize that the sample rate is five times the symbol rate. Secondly, the single-sided bandwidth of the transmit signal looks to be approximately 3000 Hz. Remembering what we have learned about pulse shaping, the required single-sided bandwidth of a digitally modulated signal at the symbol rate of 4800 symbol per second should be 2400 Hz plus the excess bandwidth from the shaping filter. In this example, the excess bandwidth is 20 percent, for a total single-sided bandwidth of 2880 Hz. So far our modulator at least passes a sanity check for the proper signal bandwidth. We will have to demodulate this signal to know for sure if everything is working properly.

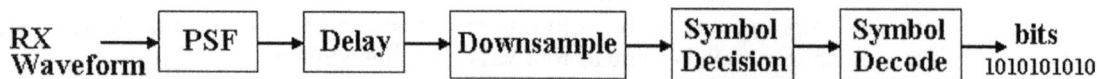

RX Waveform → **PSF** → **Delay** → **Downsample** → **Symbol Decision** → **Symbol Decode** → bits
1010101010

Figure 7.2 Flowchart showing the demodulation process for a digitally modulated waveform.

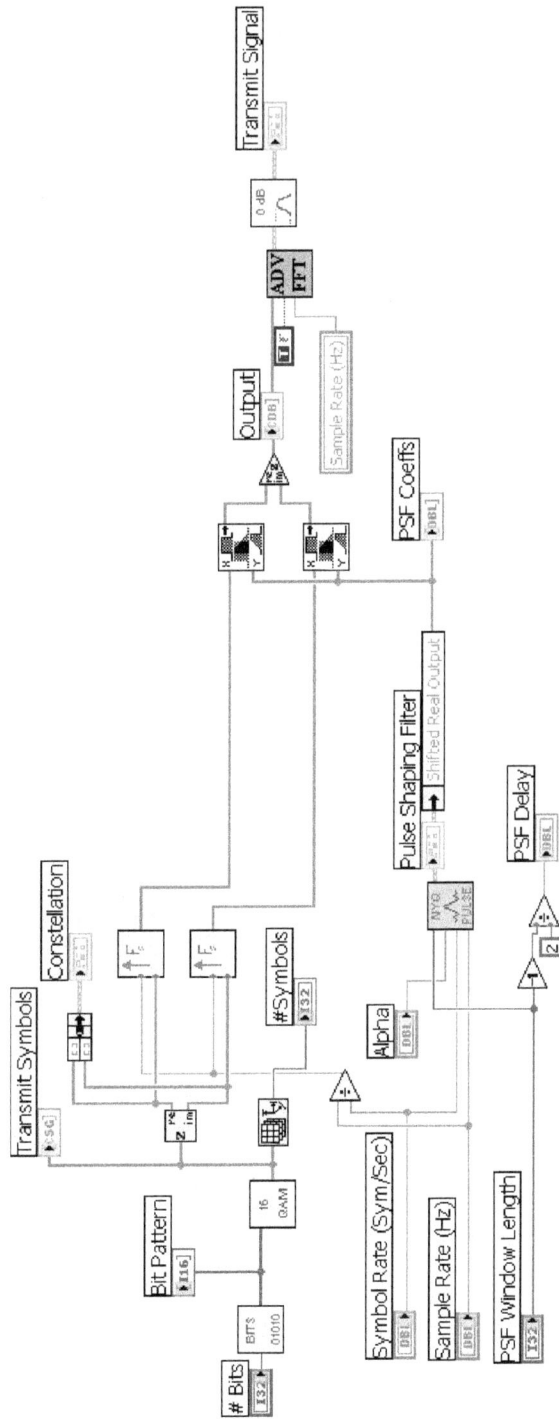

Figure 7.3 Modulator.vi block diagram.

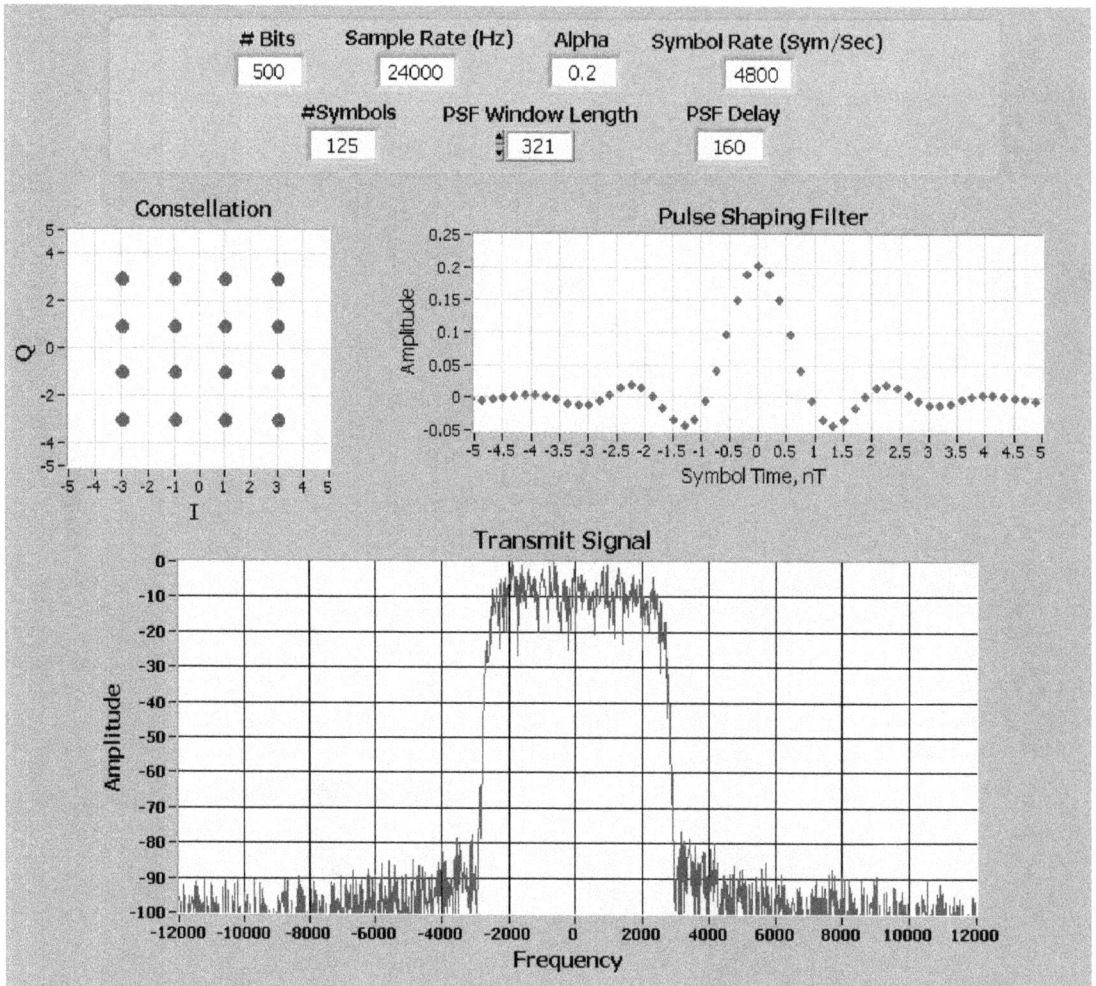

Figure 7.4 Modulator.vi front panel.

7.2 Demodulator

Before we start examining the structure of the demodulator, let us quickly look at the time-domain transmitted waveforms in Fig. 7.5 to make sure we understand what was sent.

What Fig. 7.5 tells us is that the transmitted waveform is approximately zero for some duration at the beginning and end of our transmission. Remember that the convolution output has length $L + N - 1$ where L is input waveform length (625 in this case) and N is the filter length (321 in this case). As the two waveforms slide past each other in the convolution, we get $(N - 1)/2$ zeros from the

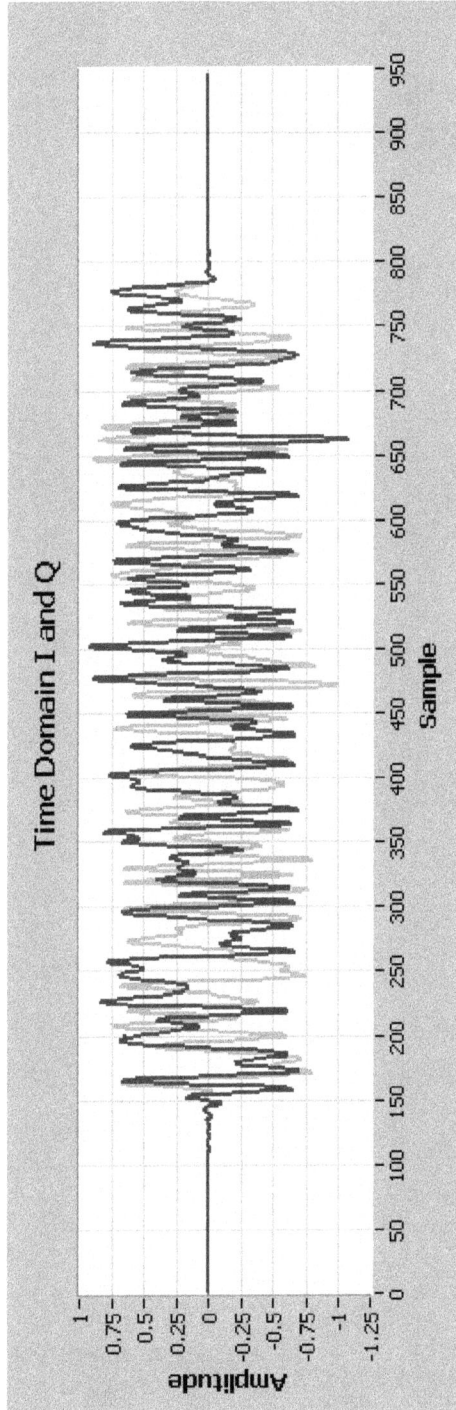

Figure 7.5 Transmitted *I* and *Q* waveforms.

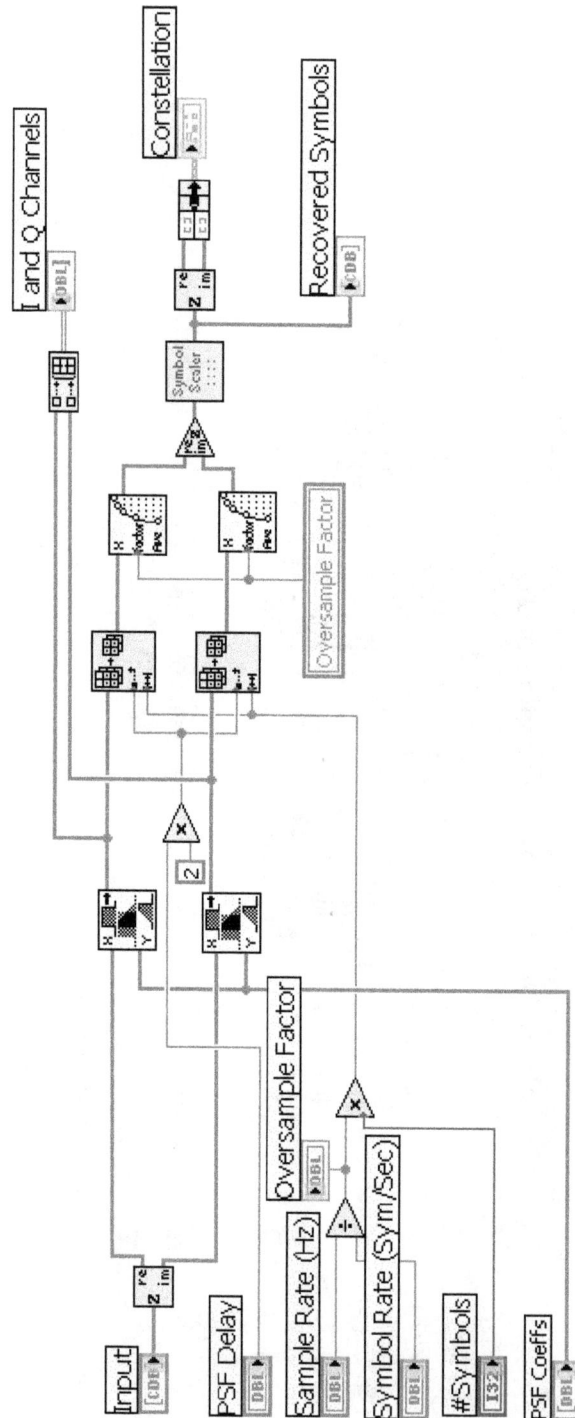

Figure 7.6 Demodulator.vi block diagram.

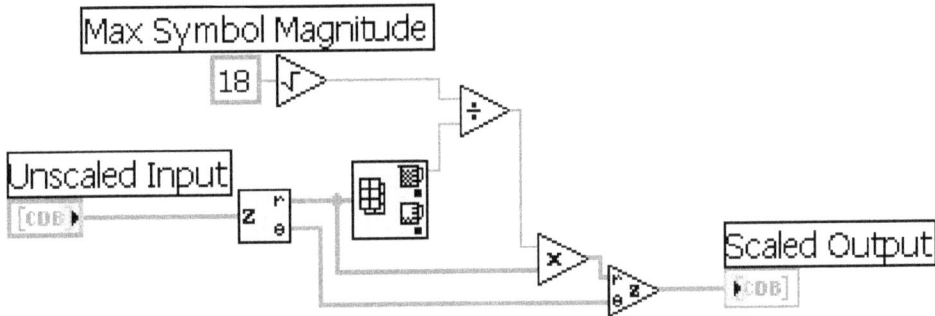

Figure 7.7 ScaleSymbols.vi block diagram.

root-raised cosine tails before the first input symbol is actually in the main lobe of the shaping filter and a reverse situation at the tail end of the waveform. That means that we have transmitted $N-1$ samples that really do not contain any information. We will get rid of them shortly once we finish the receiver half of the pulse shaping.

Following the flowchart in Fig. 7.2, the first thing the receiver needs to do is to complete the other half of the pulse-shape filtering. Remember that there is a root-raised cosine at both ends of the path to yield an overall raised cosine response. After that convolution, we have added $(N-1)/2$ more unwanted samples to the front of our waveform, bringing the total to $N-1$. For that reason, we delay the signal by $N-1$ and trim off the $N-1$ excess points at the end as well. Now we can downsample the signal to arrive at what should be the transmitted symbols. We may need to perform some scaling on the recovered symbols to account for the pulse-shaping filter gain, channel noise, and so on. Figure 7.6 shows the block diagram of Demodulator.vi, which reflects the steps discussed previously.

The VI ScaleSymbols.vi shown on the right in Fig. 7.6 takes care of the scaling mentioned previously. The block diagram for ScaleSymbols.vi is shown below in Fig. 7.7. Since we know that the magnitude of the largest transmitted symbol is $\sqrt{3^2 + 3^2} = \sqrt{18}$, this VI simply scales the largest input magnitude up (or down) to that value.

The front panel of the demodulator is shown in Fig. 7.8. You will notice right away that the recovered symbols are not quite perfect replicas of the possible constellation points, but we will fix that shortly. So far this demodulator has not made any decisions about which symbols were likely to be transmitted. In this simple example, there is no noise to corrupt our waveform so the only deviation comes from the pulse-shaping filter convolution. We could round our values to the nearest integer to clean up the recovered constellation, but instead we will utilize a decision method to choose the recovered symbol from the constellation. Before we do that though, let us look at corrupting our transmitted signal with noise.

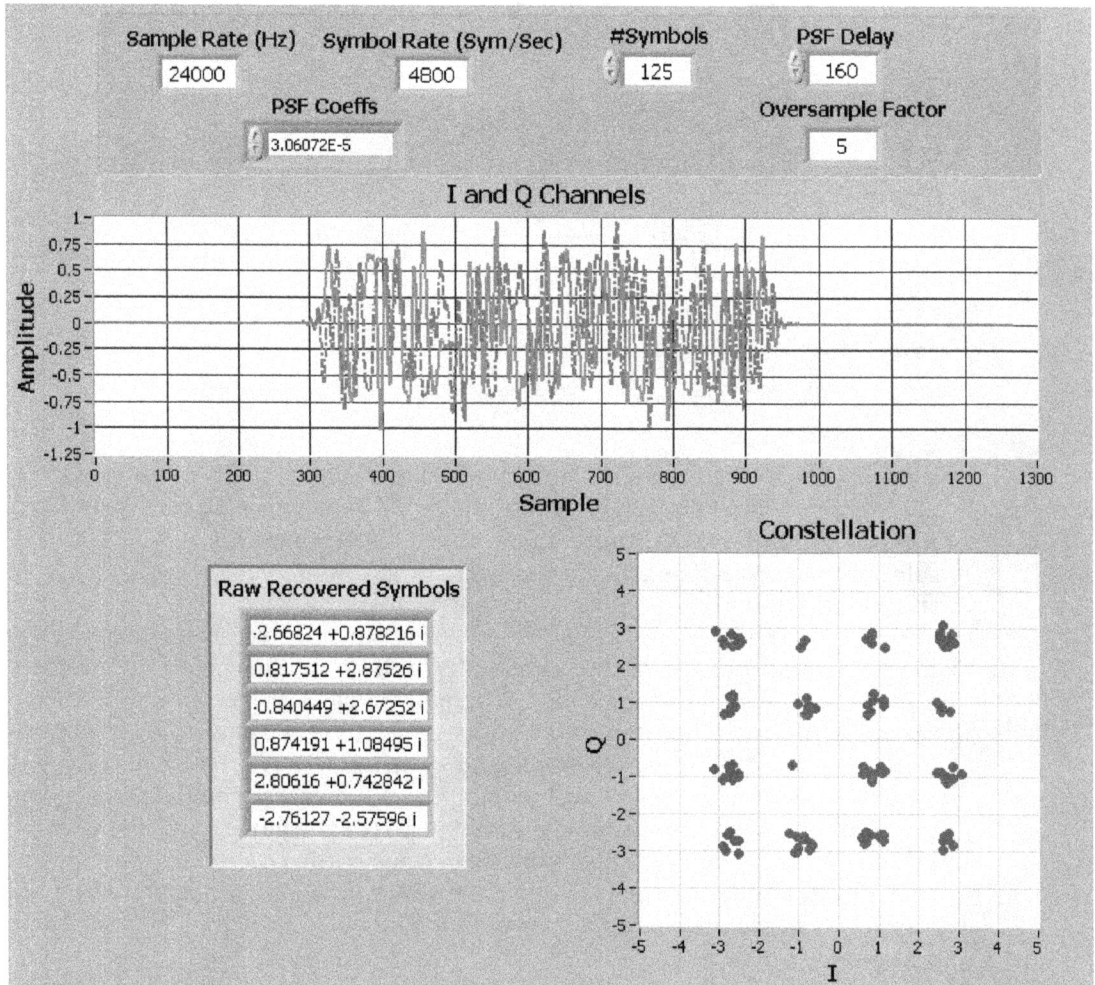

Figure 7.8   Demodulator.vi front panel.

7.3   Channel Impairments

Now we are going to start adding some degradation to the received waveform. Chapter 6 built a VI called AWGN.vi, which simply generates a white gaussian noise signal with a specified power. We use the VI as shown in the block diagram of SimpleNoisySystem.vi in Fig. 7.9. This first step is to determine our desired SNR. Once we have the SNR, we can work backward to compute the required noise power to achieve that SNR. Then we call AWGN.vi with the appropriate signal length and power and add the generated noise to our transmitted waveform.

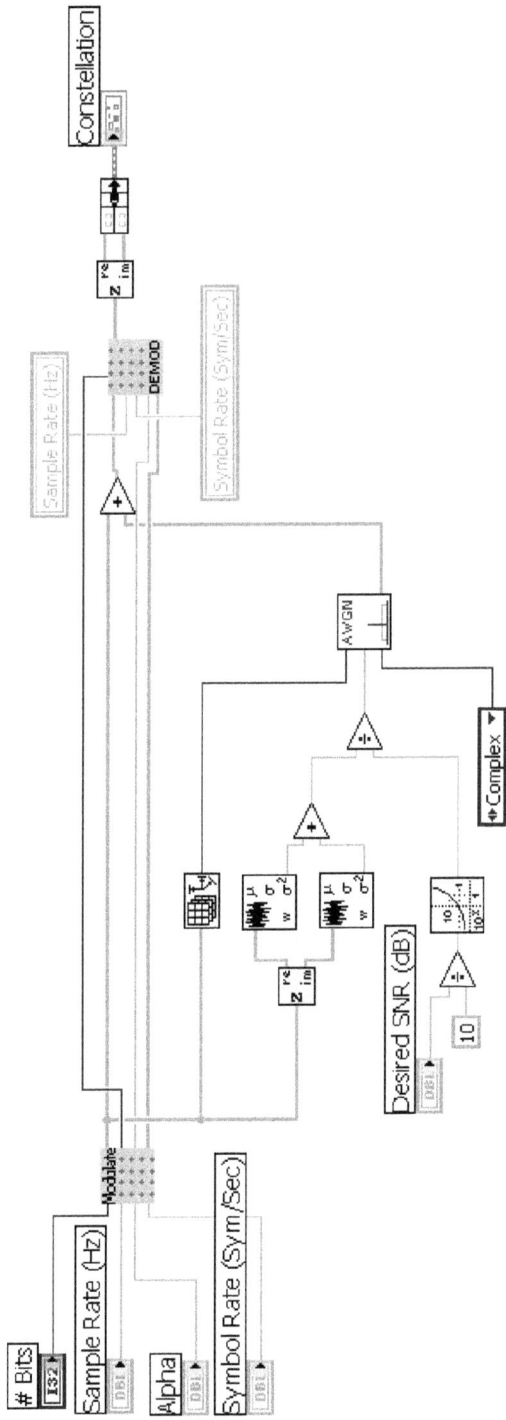

Figure 7.9 SimpleNoisySystem.vi block diagram.

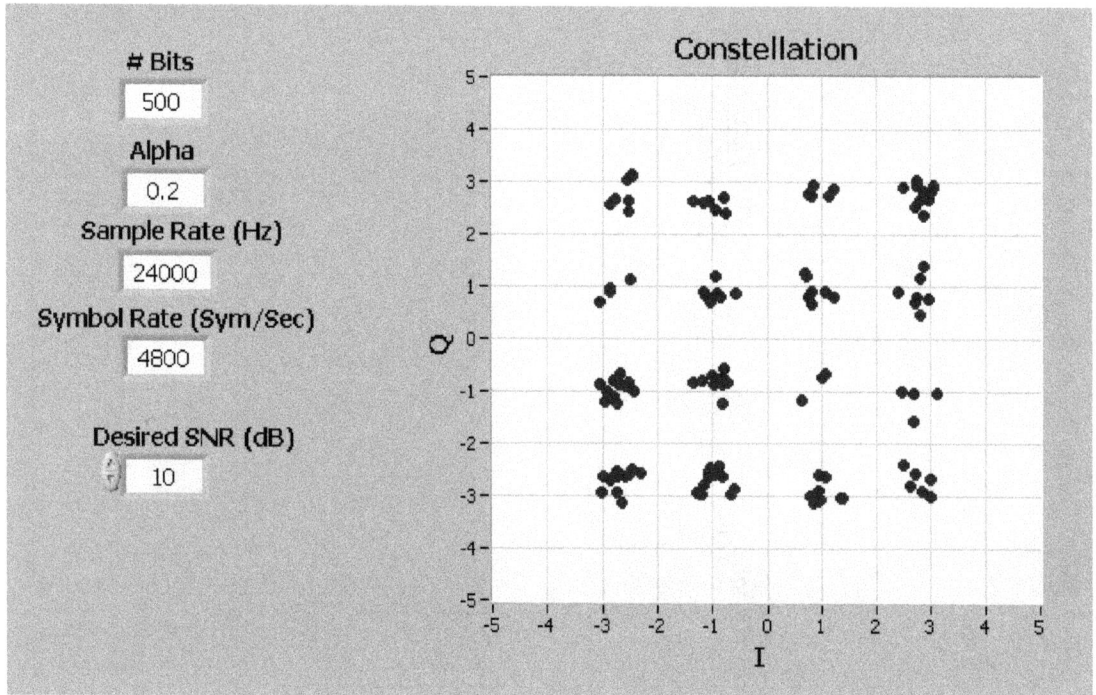

Figure 7.10 SimpleNoisySystem.vi front panel.

Figure 7.10 shows the noisy demodulated symbols. Here the SNR was set to 10 dB and it is obvious that the AWGN has spread the symbols in such a way as to increase the probability of a symbol error. By running the VI several times, you may actually see cases where a point falls on or near the decision boundary, which may cause a symbol error. Remember also that we Gray encoded the symbols such that an incorrect decision between two adjacent symbols would only yield a single-bit error.

In a conducted environment, the AWGN-induced noise may be sufficient to put our digital communication system to the test. However, for more accurate modeling of the wireless RF channel, we have to include fading. We saw in Chap. 6 that the channel could be modeled with a Rayleigh random variable and that we could sum up a sufficient number of these Rayleigh faded envelopes to model the multipath components arriving at the receiver. In fact, let us take a look at the block diagram of Channel.vi in Fig. 7.11, which applies a sum of Rayleigh faded envelopes to our digitally modulated signal.

The first step in the channel fader is to choose the number of scatterers or multipath components. Reference [2] tells us that we should be able to work with as few as six scatterers or multipath components. Then we call Rayleigh.vi (from Chap. 6) in a loop and sum up each generated complex envelope into a shift register. We choose our desired signal-to-noise ratio and calculate the noise

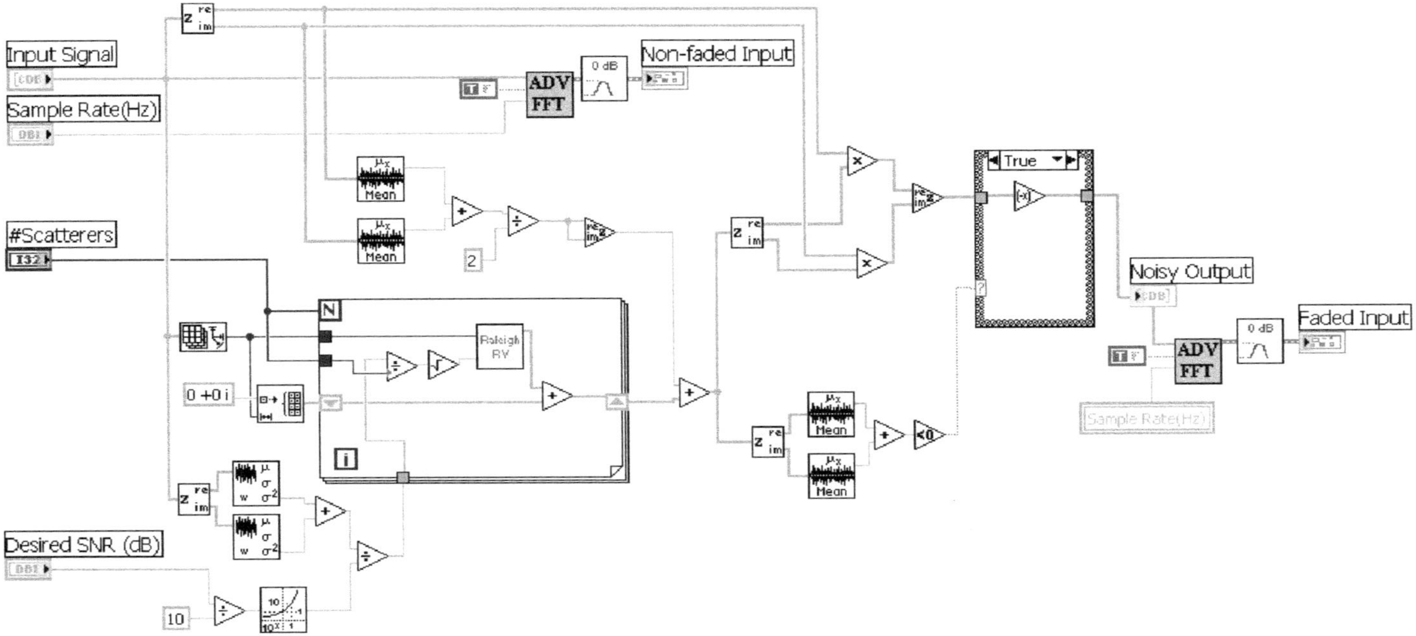

Figure 7.11 Channel.vi block diagram.

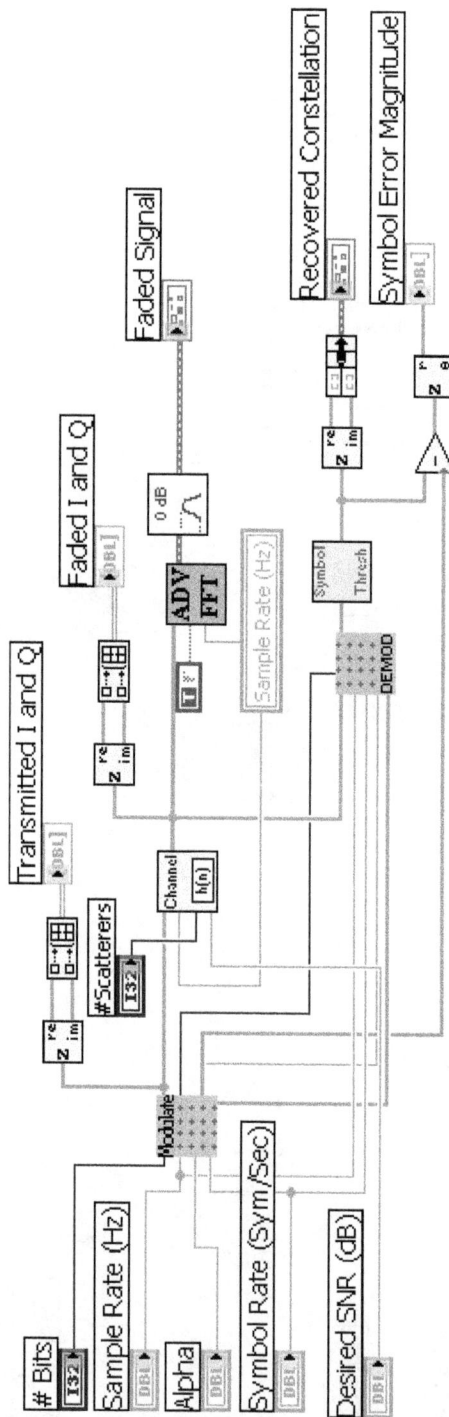

Figure 7.12 FadeSimulation.vi block diagram.

power as before. Then each Rayleigh envelope will be formed with a fraction of that noise power so that the summation yields the correct total noise power. The Rayleigh fading is a model of signal fluctuation about a mean, therefore we shift the fading envelope by the mean of the input signal and finally we can apply that faded envelope to our signal as shown in the figure.

So what is the effect of the Rayleigh fading on the received waveform? Let us take a look at a quick example of fading using Channel.vi in Fig. 7.12. We have already seen all of the VIs in this block diagram, except for the symbol detector shown just after the demodulator. This VI will be discussed in detail in Sec. 7.4.2. Its job is to make symbol decisions based on predetermined symbol boundaries.

The front panel of this simple fading simulation is shown in Fig. 7.13. The magnitude of the symbol errors are shown at the bottom of the figure. For this case of 40 dB SNR, there is a single symbol error of magnitude 2 at symbol 11. In the next chapter we will see how these symbol errors translate to bit errors.

At this point, we have only considered a static case with no motion between the transmitter and receiver. Of course, with today's ubiquitous use of wireless devices, relative motion is impossible to ignore. However, we can easily incorporate the effects of any relative motion between the base station and the mobile unit through the use of a Doppler shift of the received signal, given by Ref. [2]

$$\omega_{di} = \frac{\omega_c v}{c} \cos \psi_i \qquad (7.1)$$

where v = velocity of the mobile unit
ω_c = RF carrier frequency
c = speed of light
ψ_i = uniformly distributed angles of arrival of the i reflected waves

The effect of the Doppler shift can then be applied to the complex envelope $g(t)$, giving

$$g_D(t) = g(t)e^{\omega dt} \qquad (7.2)$$

From here it is left up to the reader to apply the effect of a Doppler shift to the digitally modulated signal. The uniformly distributed angles can be easily generated and the carrier frequency and velocity are specific to your application. Again this channel model is by no means complete. By building only the small-scale Rayleigh faded envelope, we have neglected the effects of path loss and there are no provisions for a line-of-sight component, both of which are important pieces of a full-blown channel model. However, what we have shown is that we can generate a channel model with some fairly simple building blocks in LabVIEW.

7.4 Signal Detection and Recovery

Finally we have come to the point where we can attempt to recover our transmitted waveform. We have (hopefully) eliminated ISI, modeled the Rayleigh fading, and even threw in some gaussian noise. At this point, you might think

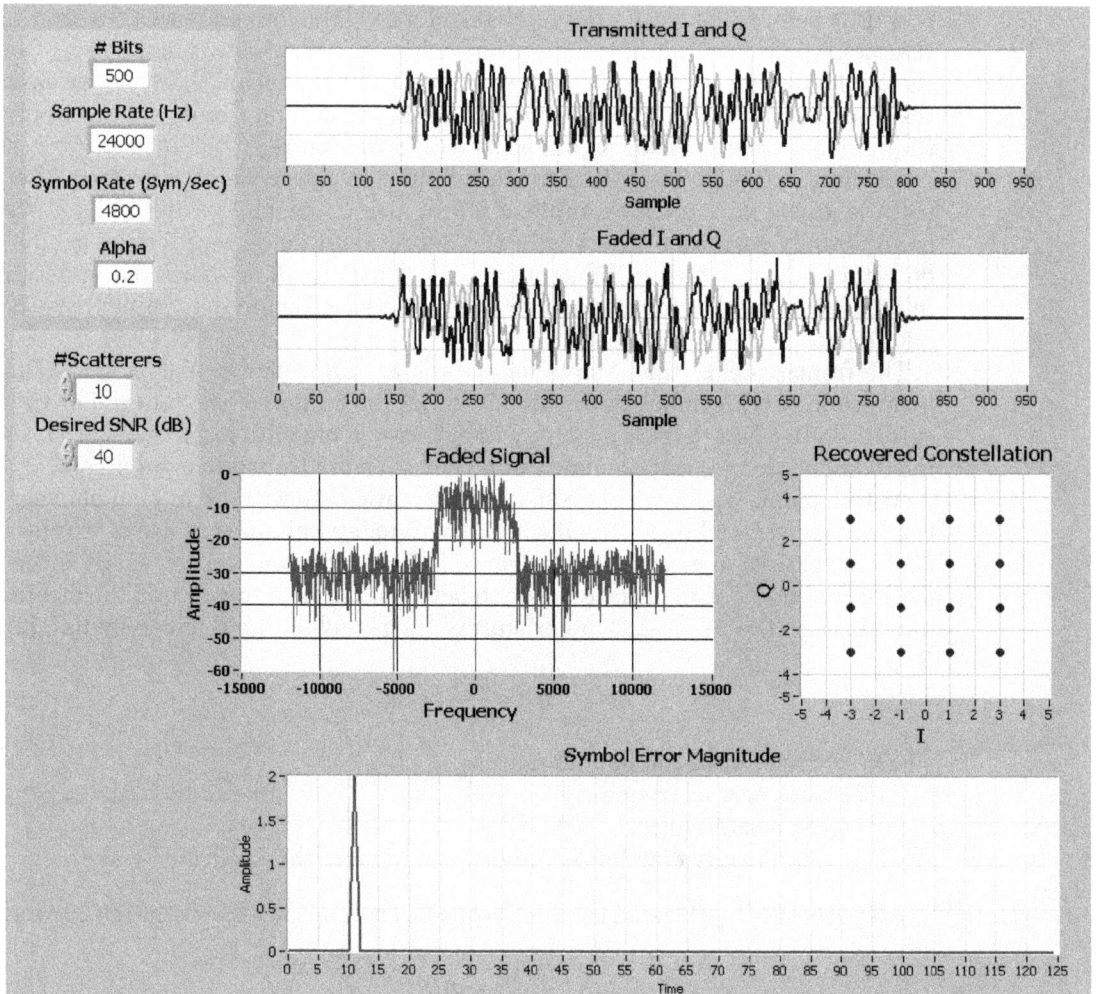

Figure 7.13 FadeSimulation.vi front panel.

our signal is unrecoverable, but do not worry, our signal is still in there somewhere. Our first step on this end of the chain will be to filter off some of the noise from the channel and any interferers that may be out there. We have to be careful not to have too narrow a filter bandwidth because the frequency error (if there is any) may just push our signal outside the filter bandwidth.

Of course, in general, there will be some receiver front-end operations that must happen in order to get to the baseband waveform that we are dealing with here. Those operations include analog-to-digital (A/D) conversion, mixing the RF signal possibly to an intermediate frequency (IF) before the final downconversion to baseband, and possibly several stages of resampling. The details of the A/D conversion were discussed in Chap. 2 and will depend on your specific hardware capabilities. If the downconverted waveform is sitting at some frequency other than dc, the function ComplexMixer.vi can be used to mix the signal the rest of the way to dc.

7.4.1 Matched filter detection

The topic of matched filters might be just a little misleading. This is not a filter in the traditional sense. Normally a filter is concerned with removing some portion of the spectrum that is unwanted and the remaining part of the spectrum is passed as intact as possible. In general, the goal of a filter is to preserve the desired signal with as little distortion from the filter as possible. A matched filter on the other hand, does none of that. In fact, the output looks nothing like the desired signal [3]. The purpose of a matched filter is simply to compute a metric to help us decide whether or not a signal is present. In a moment we will see how a matched filter can be implemented in LabVIEW.

Suppose you were shown the noisy waveform at the bottom of Fig. 7.14 and asked to decide whether the chirp signal with $k = 2$ or with $k = 3$ were present. With some very close inspection, you might be able to say which one was sent. You would probably try to match up the peaks in the noisy signal to peaks in each of the known signals and see which one is the closest. We can accomplish almost the same thing with the matched filter. We know beforehand that the noisy signal can only be one of two signals. Therefore, if we use those two known signals as a template, we can correlate the known templates with the noisy signal and whichever correlation is the highest is most likely the signal that was transmitted.

In the block diagram shown in Fig. 7.15, the chirp signal for $k = 2$ is corrupted with AWGN noise. If the signal set is assumed to contain only a chirp for $k = 2$ and for $k = 3$, then we can simply use those two signals as a template for the correlations. The output of the correlation is shown in Fig. 7.16.

It is immediately obvious from the matched filter outputs that the $k = 2$ chirp is the signal that was sent. This simple two-signal example can be easily extended to the 16-QAM case with some minor changes. It is also a good idea to normalize the correlation since the amplitudes of the 16-QAM constellation points vary. That amplitude variation could cause false results from the correlation. A normalized correlation can be computed from the following formula:

$$\text{Norm correlation} = \frac{\Sigma xy}{\Sigma |x| |y|} \tag{7.3}$$

7.4.2 Threshold decisions

Another very simple way to make a symbol decision is to simply set decision boundaries. For instance, in our 16-QAM case, any I or Q value +2 or greater maps to a +3, anything between 0 and +2 maps to +1, and so on. Let us take a look at a very simple hard decoder in LabVIEW shown in Fig. 7.17. This is the block diagram of Symbol_threshold_det.vi and we saw that it was used back in Fig. 7.12 in FadeSimulation.vi. This VI consists of a loop over the raw recovered symbols and a series of case selectors for the I and Q channels. You can see in the bottom case selector, any Q symbol +2 or greater is set to +3.

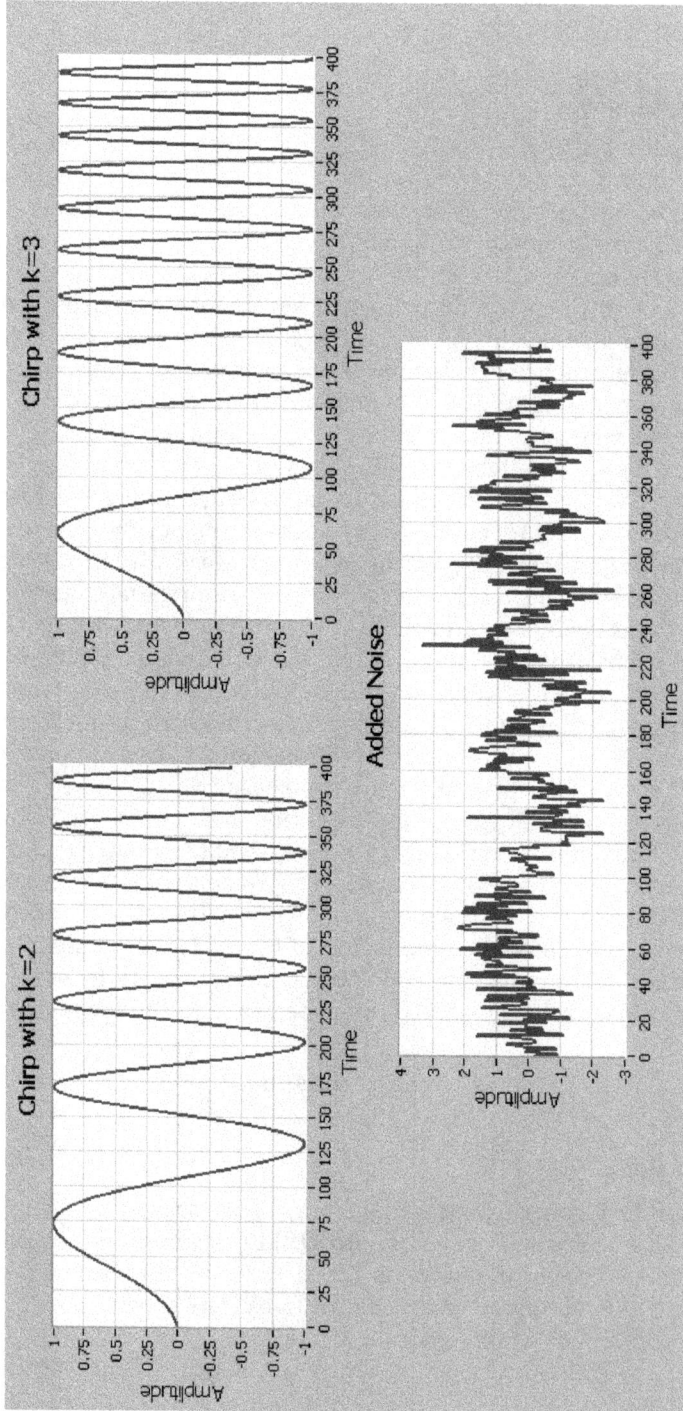

Figure 7.14 MatchedFilterDetection.vi front panel.

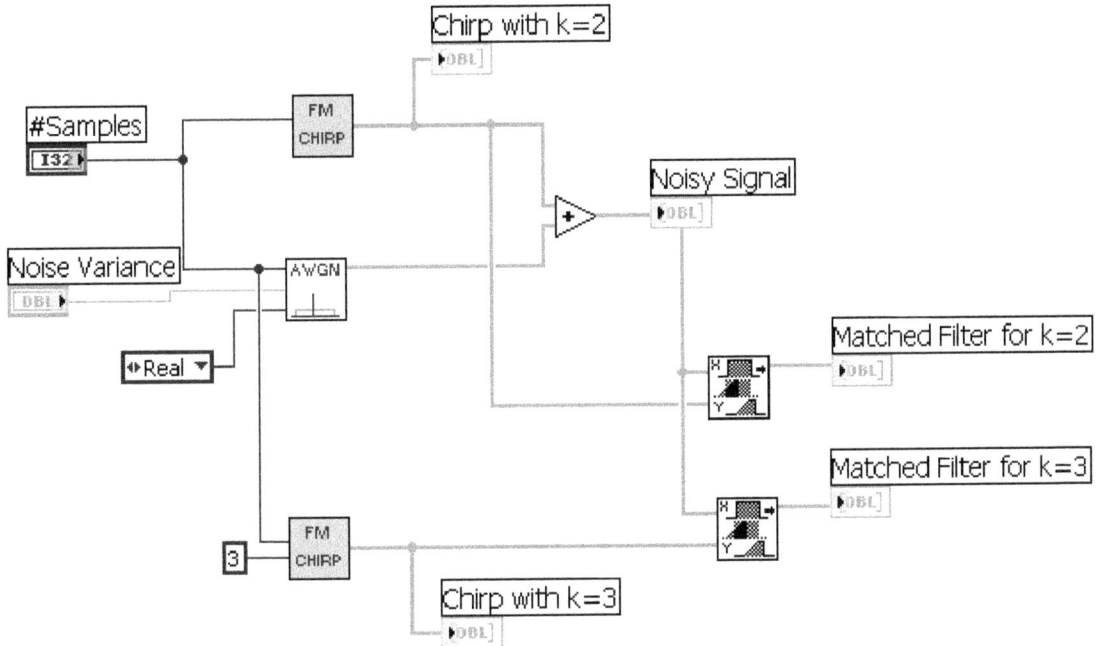

Figure 7.15 MatchedFilterDetection.vi block diagram.

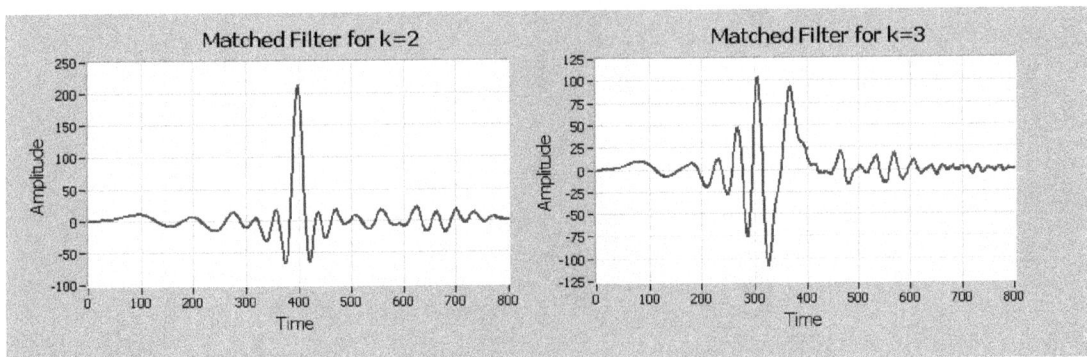

Figure 7.16 Matched filter outputs.

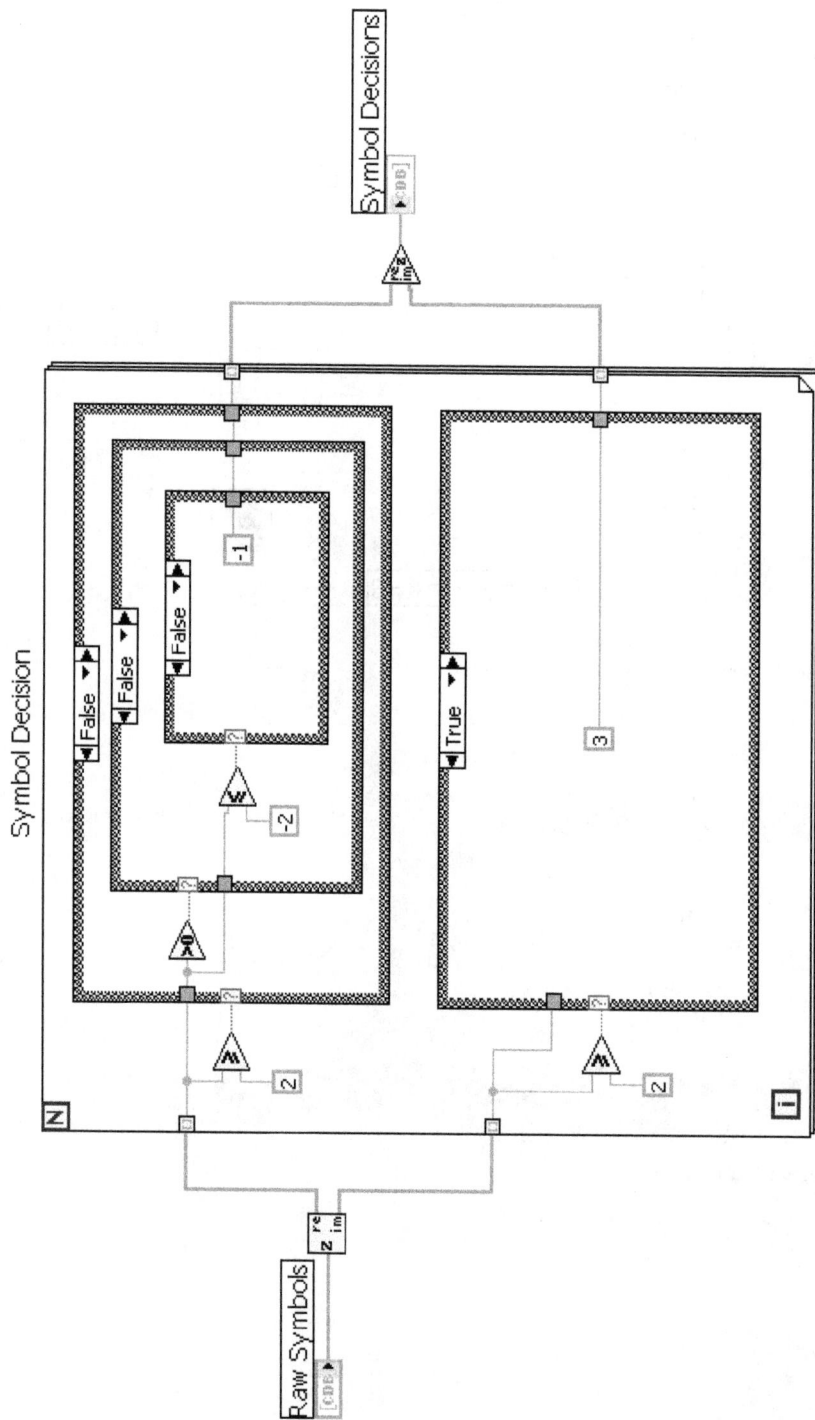

Figure 7.17 Symbol_threshold_det.vi block diagram.

7.5   Synchronization

As we are about to find out, synchronization is an extremely important part of correctly recovering a digitally modulated signal. The next two sections will discuss the important aspects of both time and frequency synchronization and how synchronization affects our ability to recover a signal. Most digital standards already include provisions for timing and frequency correction, now we will see why.

7.5.1   Time synchronization

With the waveforms we have seen so far, there have been no timing issues. We inherently knew where the waveform begins and where to start the pulse-shaping filter. But how can this be accomplished blindly, as would be the case if the receiver started receiving somewhere in the middle of a transmission? Would the receiver ever be able to recover, and if so how would it know which sample is the beginning of transmission? Obviously this is a huge issue in the world of digital communications, because if we do not know which symbol was the beginning of a transmission, there is no way the rest of the bits will be properly recovered. Actually we can apply what we have already learned about matched filter detection to the problem of time synchronization.

The output of the matched filter was a measure of the likelihood of the presence of a given signal. If we are clever, we can embed a signal, any signal we choose, at the beginning of a transmission. Well, actually the signal had better not be the same as any of our data signals or else we will get false matches. So we should choose a very distinct signal—one that has almost no correlation with any of our data symbols and just append that signal to the beginning of every transmission. Then we can use a matched filter (matched to this distinct signal) at the receiver. The peak of the matched filter output should mark the position of our embedded synchronization symbol and thus the beginning of our transmission. Also this is another case where normalizing the correlation using Eq. (7.3) will be extremely important to avoid false detections from any large amplitude spurs in the received signal.

7.5.2   Frequency synchronization

Correct frequency alignment is just as important as the time synchronization mentioned previously. The effect of a small frequency error of 10 Hz on the signal constellation is shown in Fig. 7.18. What you see in the constellation is concentric rings. Since the units of frequency are radians per second, any frequency error will impart a phase rotation on the recovered symbols, which accumulates with increasing time and causes the rotating effect.

Just like the time synchronization case, to estimate the frequency error we can embed a specific symbol at known places in our transmission. These symbols will have known phases (as opposed to random data symbols, which have unknown phases) and those phases can be used as references to calculate the phase error of the recovered symbols. Exactly where in the transmission these symbols are located varies depending on the standard (GSM, APCO25, and

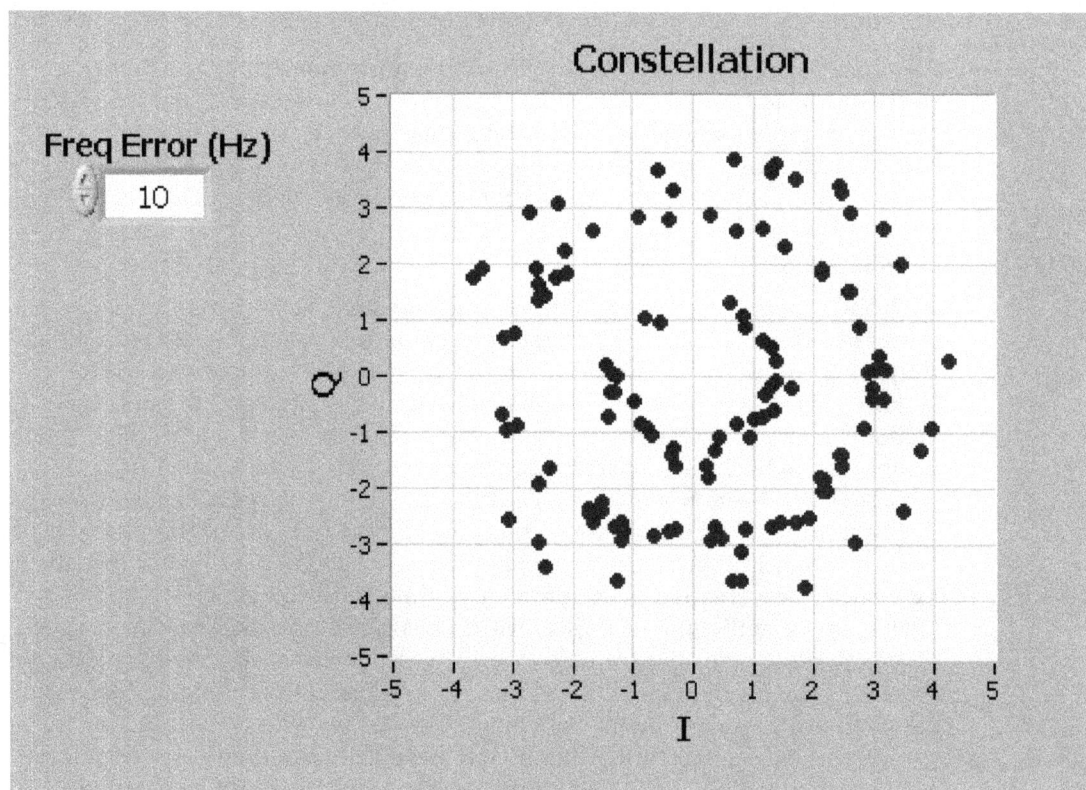

Figure 7.18   Effect of frequency error on recovered symbols.

some others), but typically there are a few of these "pilot" symbols spread throughout the transmission interval. By measuring the accumulation of phase error over the duration of the transmission, a good estimate of frequency error can be calculated.

7.6   NI Modulation Toolset

At this point it is worth mentioning that much like the Matlab toolboxes, National Instruments (NI) also has some add-on digital communication tools. We saw earlier that NI has the spectral measurement toolset, but I would characterize that package as more signal acquisition oriented. On the other hand, the modulation toolset contains several very useful functions for measuring and analyzing communication signals, some of which are briefly mentioned here. At present, this toolset is only available with the purchase of the PXI-5670 vector signal generator, but hopefully in the future this toolset will be available separately. The typical location for the toolset is shown in Fig. 7.19.

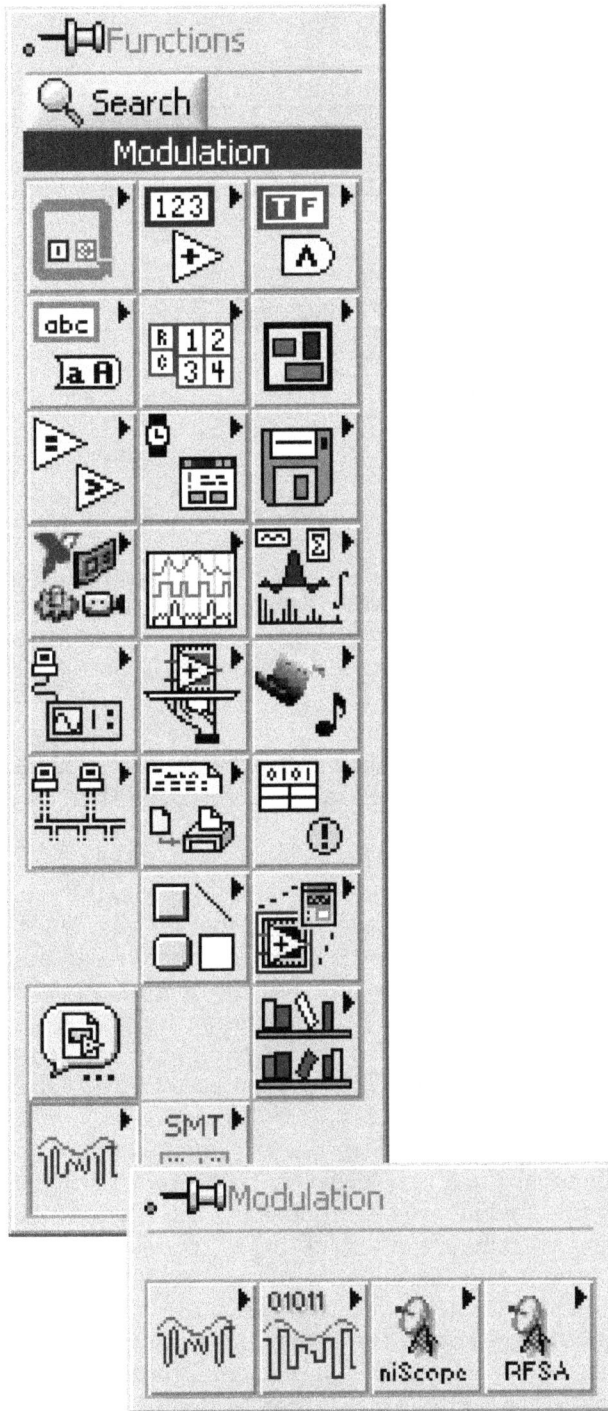

Figure 7.19 Location of modulation toolset palette.

Analog Modulation Tools

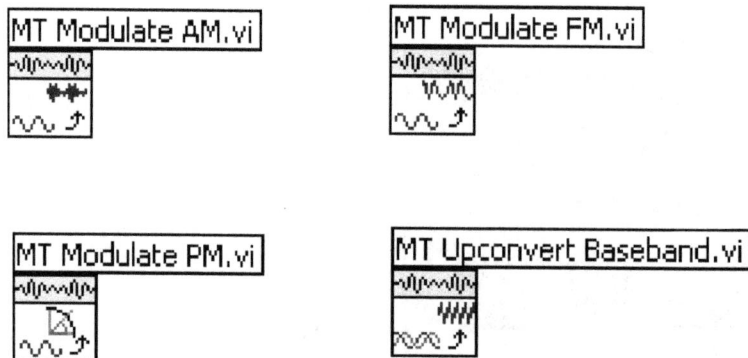

Figure 7.20   Modulation toolset analog tools.

The modulation toolset comes with analog and digital modulation tools as well as some utility functions to perform the auxiliary operations that go along with digital communications. Here we will only mention a few of these functions to give the reader an idea of what the toolset capabilities are, for more information check out *www.ni.com*.

Let us start with the analog modulation tools. The four analog functions that are included with the toolset are shown in Fig. 7.20. These functions will generate amplitude, frequency, and phase modulation and there is also one VI to upconvert the baseband signal.

A few of the more interesting digital modulation tools are shown in Fig. 7.21. In the light of what we have seen previously in this book, these particular tools are the most relevant. These functions will generate bits, QAM symbols, add AWGN noise, apply fading, and even perform fractional resampling. Not shown are several other functions for alternate forms of modulation such as FSK and MSK as well as the corresponding demodulation functions. Of these digital modulations functions, the MT Fraction Resample.vi has proven to be very useful in some of my own work. This function performs an efficient resample and filter operation. The user needs only to supply the waveform to be resampled, the current sample rate, and the desired sample rate.

NI's primary goal for these tools seems to be to promote the use of their PXI RF hardware, some of which was mentioned back in Chap. 2. This section has not gone into detail on the use of any of these functions; however, the hope is that the information in this book should give the reader a good understanding of what is going on behind the scenes in each of them. The NI help will also give you a good understanding of how these functions were intended to be used. The fact that these functions exist is a good indication that NI is seeing a demand for their products in the digital communication arena and hopefully there will be more of a push to include these tools with the standard LabVIEW package.

Digital Modulation Tools

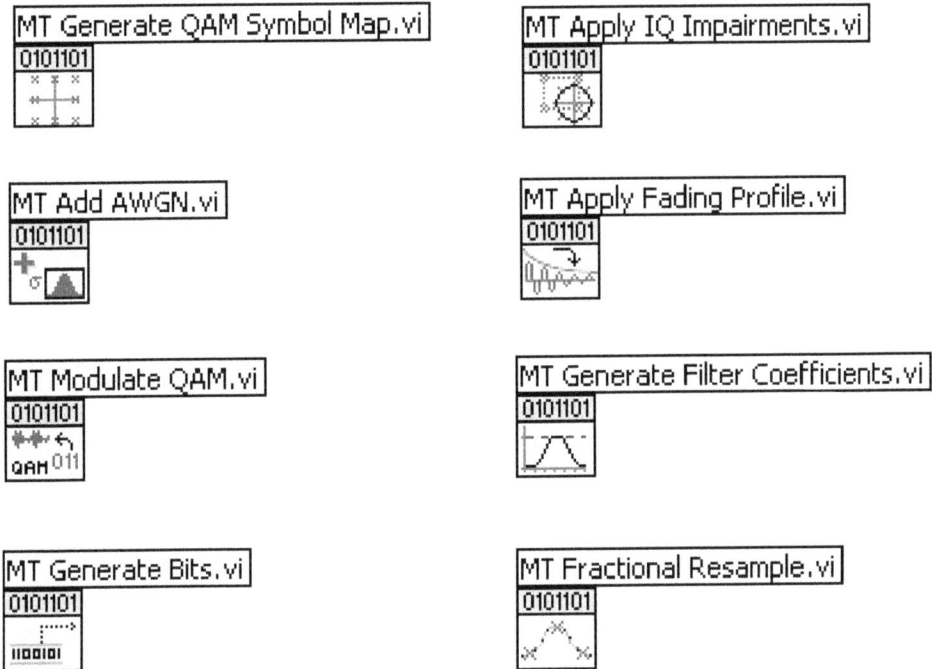

Figure 7.21 Modulation toolset digital tools.

Summary

Chapter 7 has finally brought everything together into a digital communication system. We started by assembling the pieces of the modulator and the demodulator. Then we added channel impairments such as additive white gaussian noise and Rayleigh fading and we saw the effect of those impairments on the recovered symbols. Next we examined some methods for making symbol decisions. We also discussed the importance of synchronization in both time and frequency. Finally we ended with a brief overview of the NI modulation toolset.

References

1. Oppenheim, A. V., R. W. Schafer, and J. R. Buck, *Discrete-Time Signal Processing,* 2d ed., Prentice-Hall, Upper Saddle River, NJ, 1998.
2. Prabhu, G. S., and P. M. Shankar, "Simulation of Flat Fading Using MATLAB for Classroom Instruction," IEEE Trans. on Education, vol. 45, no. 2, February 2002, pp. 19–25.
3. Sklar, B., *Digital Communications,* 2d ed., Prentice-Hall, Upper Saddle River, NJ, 2001.

System Performance

So far we have seen that LabVIEW really can do quite a bit of signal processing and we have used those capabilities to simulate a whole digital communication system. Now it is time to investigate what measures of performance we can do to see how our communication system stacks up. But what measurements will tell us if our system is good or not? Since the ultimate goal of a digital communication system is to convey a bit sequence from point A to point B, the first measurement that stands out is the number of bit errors. In Chap. 7 we decoded our recovered symbols into their corresponding bits, so if we know what bits were sent, we can determine how many errors we have. However, the number of bit errors (or the bit error rate) may not tell the whole story of imperfections in the recovered waveform. In fact, for some constellations even large recovered symbol deviations may not result in even a single bit error. For those cases, it will be necessary to look at other signal metrics. Finally, we will see that there are a few techniques to give our signal a boost of immunity to channel degradation.

8.1 Performance Measurements

8.1.1 Bit-error rate

If we want to evaluate the error performance of the digital communication system, counting bit errors certainly seems like the right place to start. In Chap. 7, we saw some techniques for recovering the transmitted symbols, now we need to unmap those symbols back into bits. The block diagram in Fig. 8.1 shows basically the reverse process from the SymbolMapper.vi in Chap. 6. Here the same Gray coded symbol table is searched for the correct recovered symbols. Next the index of the correct match is converted to a boolean array and then cut down to four elements (for the 16-QAM case). Then the flipped boolean array is converted to binary and the resulting decoded bits are appended to the output bit array.

Of course to be able to determine which bits are in error, we have to know the exact sequence of transmitted bits. Take a look at Fig. 8.2 where the SymbolDecoder VI

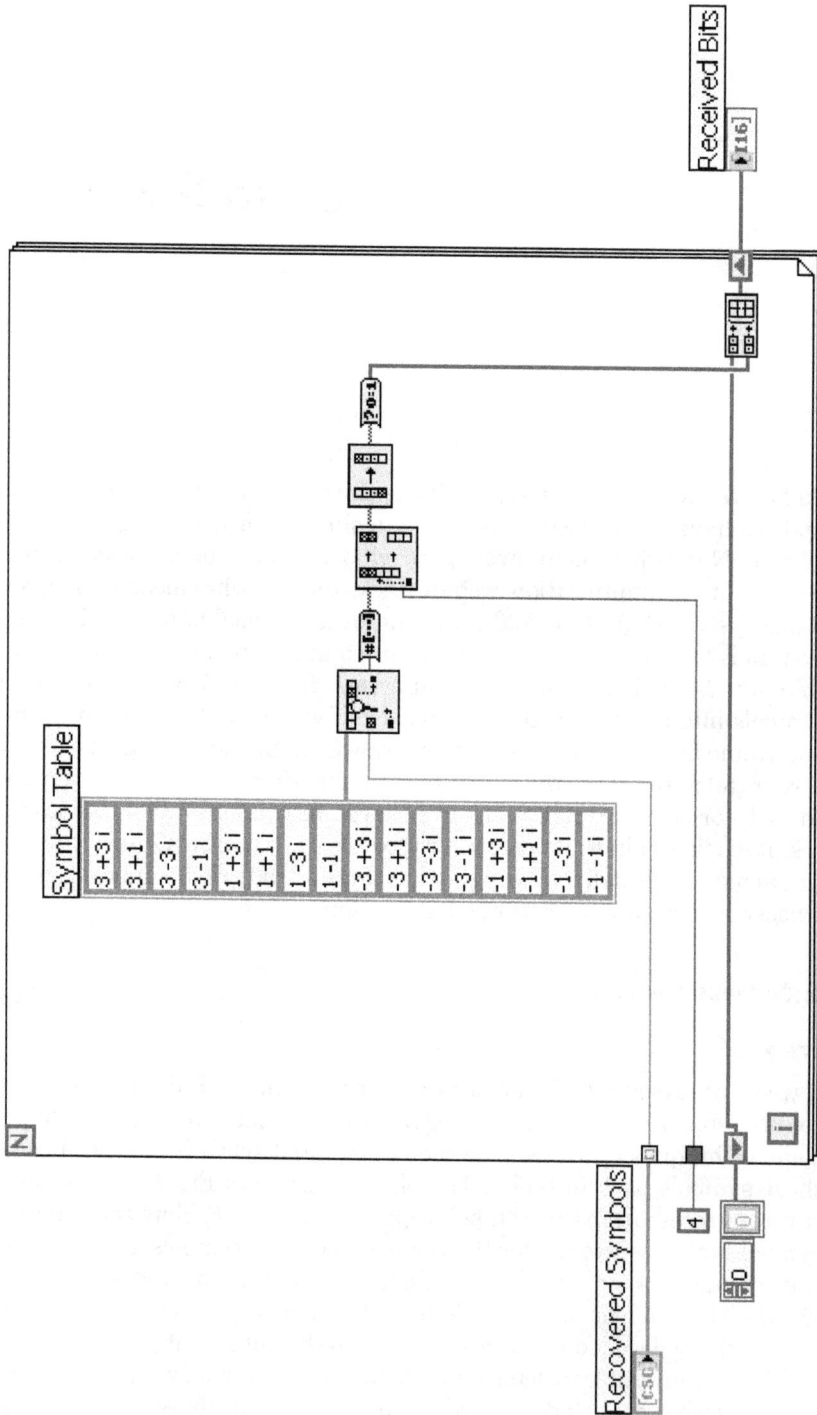

Figure 8.1 SymbolDecoder.vi block diagram.

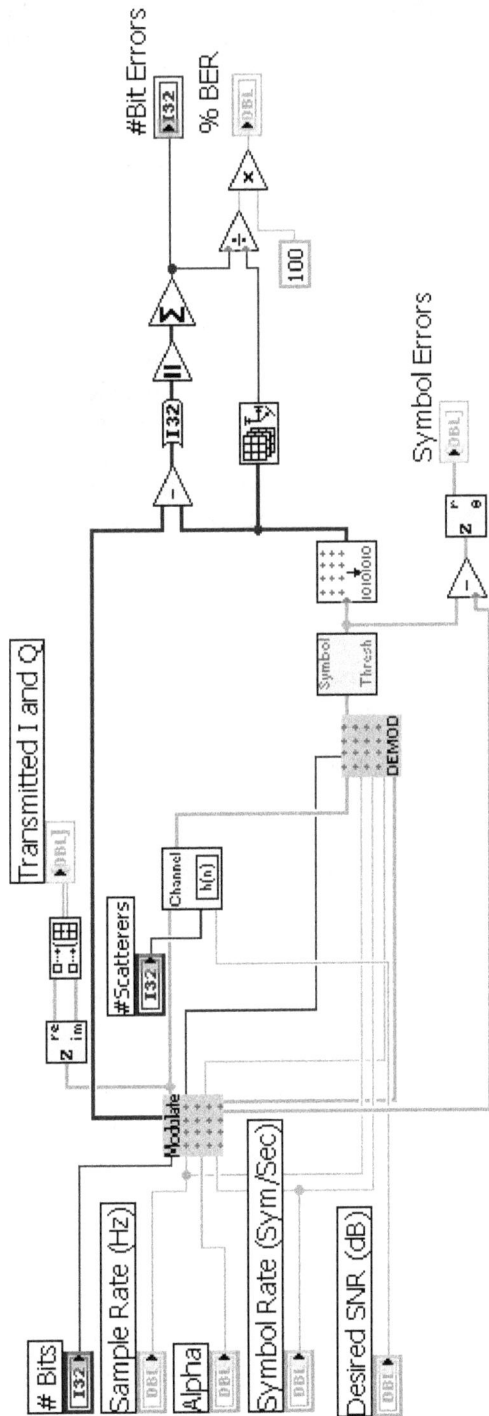

Figure 8.2 BERTest.vi block diagram.

is used and the number of bits in error is determined by subtracting the recovered bit sequence from the known bit sequence and summing up the result.

If you count the symbol errors in Fig. 8.3, you will see that there are seven places where the transmitted and received symbols do not agree. At the same time, there are only seven bit errors in the decoded symbols, which is a direct result of the Gray coding. Without Gray coding, a symbol incorrectly decoded near the decision boundary could have up to 4 bits in error (assuming 16-QAM), but the Gray coding has limited us to 1-bit differences for neighboring symbols.

When we looked at the AWGN corrupted 16-QAM constellation back in Fig. 7.10, it may have occurred to you that in some cases it takes a lot of noise to actually cause a symbol error and subsequent bit error. In other words, the raw received symbols may be pretty far from their ideal location and still not cause an error. For those cases, the bit error rate really does not tell us much about the quality of our signal, especially for very sparse constellations such as QPSK where symbol decision regions are large. Next we will look at some metrics that examine more the quality of our received signal without regard to the bit errors.

8.1.2 Error vector magnitude

This measurement tells us how much our radio frequency (RF) hardware and the communication channel have corrupted our recovered symbols. Error vector magnitude or EVM is a simple measure that relates to the distance from the actual recovered symbol to the corresponding constellation point. Reference [1] says that EVM is normally reported as a percentage of the peak signal level where the peak is defined as the constellation corner state. For our 16-QAM constellation, the peak magnitude is $\sqrt{18}$ and that value is used to normalize the magnitudes of the error vectors in the block diagram of EVM.vi shown in Fig. 8.4. The actual percentage of EVM for each transmission is taken as the mean overall of the recovered symbols.

Figure 8.5 shows the front panel of our EVM function. For the received constellation shown, the average percentage of EVM is calculated to be 8.2 percent. This particular received constellation will probably have no symbol errors (or bit errors), but measuring the EVM will give us a way to analyze and compare the symbol corruption in spite of the lack of symbol errors.

8.2 Improving System Performance

We have just seen a couple of measurements for evaluating how well our communication system processes our waveforms, but what can we do if the readings are not good? There are many established methods for dealing with channel impairments such as diversity, equalization, channel coding, and channel estimation. The next sections deal with a few of these topics.

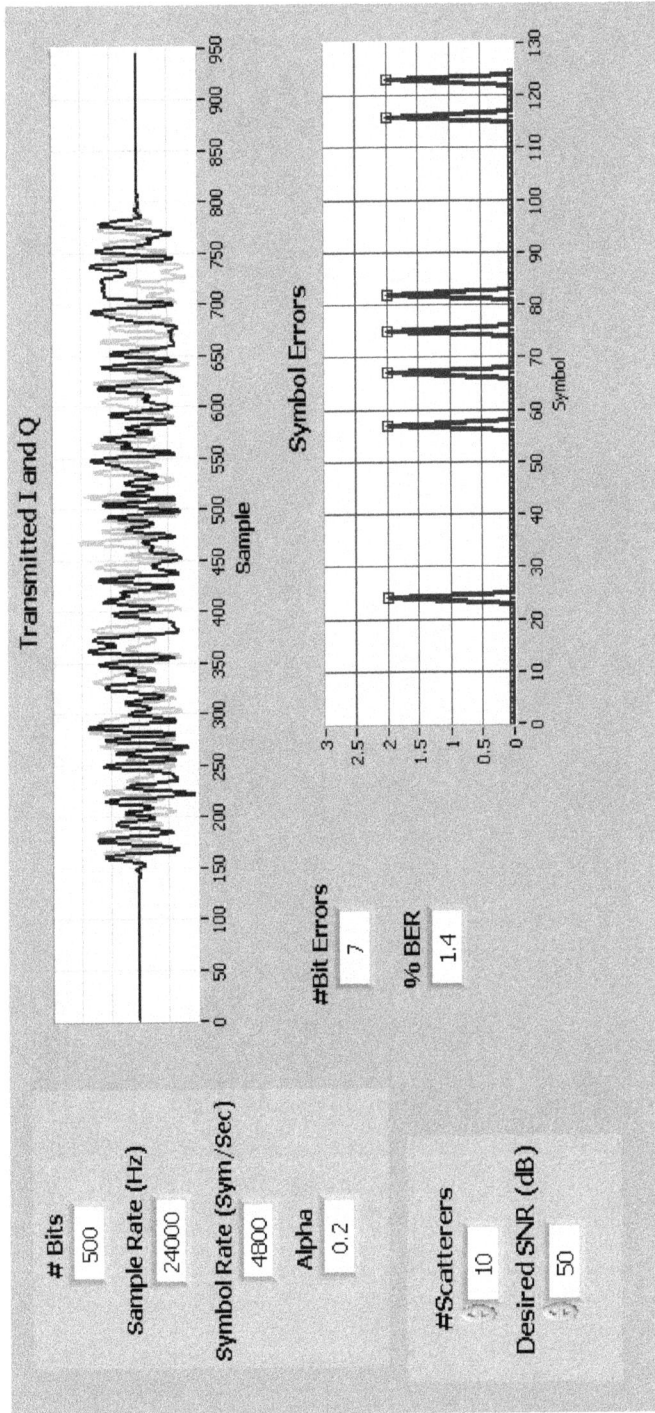

Figure 8.3 BERTest.vi front panel.

Reference Symbols

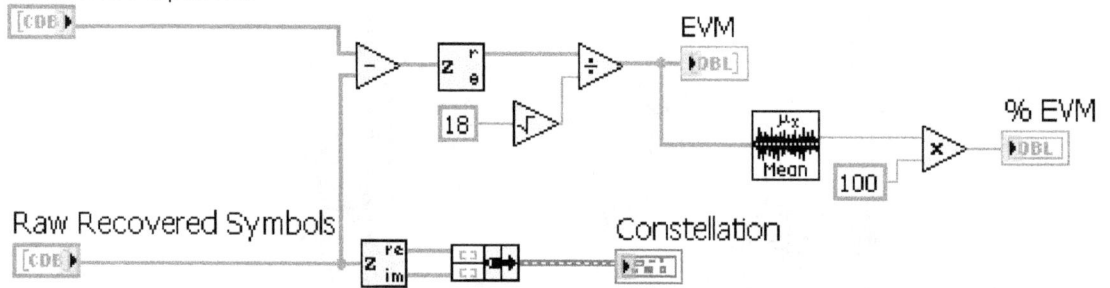

Figure 8.4 EVM.vi block diagram.

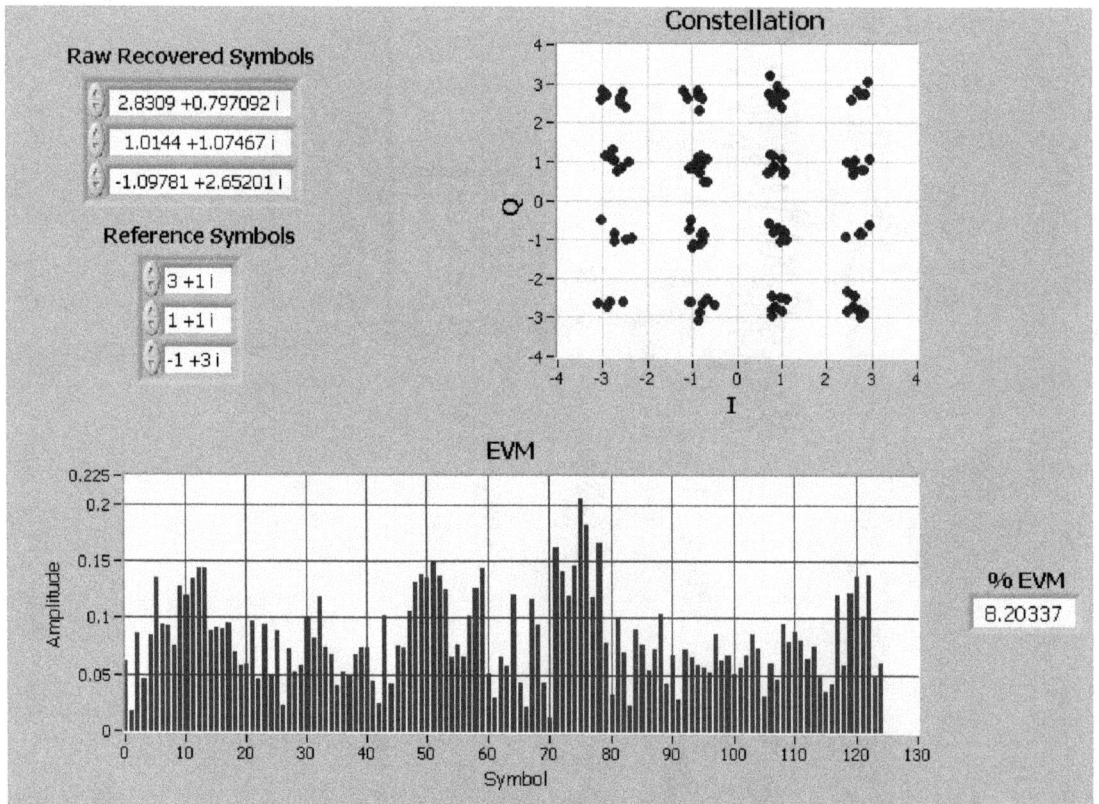

Figure 8.5 EVM.vi front panel.

8.2.1 Channel estimation

When combating the effects of any wireless RF channel, estimation is a good place to start. Channel estimation is an educated guess as to the amplitude and phase corruption imparted on our signal by the channel. A very simple way to estimate this corruption is to embed a known signal at a predetermined point in the transmitted waveform. For instance, at the end of the transmission we could append a certain symbol. We could then compare the recovered symbol's amplitude and phase to the reference and thereby deduce the channel properties. Hopefully by correcting the damage caused by the channel, we can improve both the EVM and the bit error rate (BER).

Figure 8.6 is basically the same demodulator block diagram we have seen before. The difference is that the complex mixer is used to introduce some frequency error into the recovered symbol waveform. Also not shown is the fact that the modulator split the output symbol stream in half and inserted a $2 + j4$ symbol as a pilot in the middle of the data (at index 63). This symbol is clearly not one of the 16-QAM symbol points and the demodulator removes it from the data symbols with the split 1-D arrays shown in Fig. 8.6. Once the pilot symbol is extracted, its phase is compared to the reference phase from the $2 + j4$ symbol. From that comparison, the resultant phase error is then used to calculate an estimate of the frequency error for the data symbols.

In Fig. 8.7 we see the front panel of the modified demodulator mentioned previously. The concentric rings characteristic of frequency error are clearly visible and it is obvious that the demodulated constellation will result in large bit errors. We have the luxury here of knowing that the mixer frequency was set to 20 Hz so we can immediately see that the pilot symbol accurately measured the frequency error. Knowing this error, we could apply another mixer at –20 Hz to the recovered symbols.

Of course there is much more to channel estimation than this simple example reveals. For instance, we could also use the pilot symbol to correct any amplitude corruption of the data symbols. And for best results, we would embed more than one pilot evenly spaced throughout the transmission.

8.2.2 Channel coding

Channel coding refers to a method of insulating the transmitted data from errors by introducing redundancy in the bit sequence. Typically the redundancy increases the length of the transmitted message (by adding code bits) so either the data rate or constellation density must change in order to maintain the uncoded bit rate. There are many different types of coding and a thorough treatment can be found in Refs. [2] and [3]. One special coding scheme known as trellis-coded modulation (TCM), combines coding and modulation into a single operation. The interesting feature of TCM is that the data rate is not increased, nor is the bandwidth or required signal power [3]. So do we get something for nothing here? Not quite, the real cost of TCM is on the receiver end in the complexity of the decoding process. We examine the decoding process in the next section, but first let us look at how to design a trellis-coded modulator in LabVIEW.

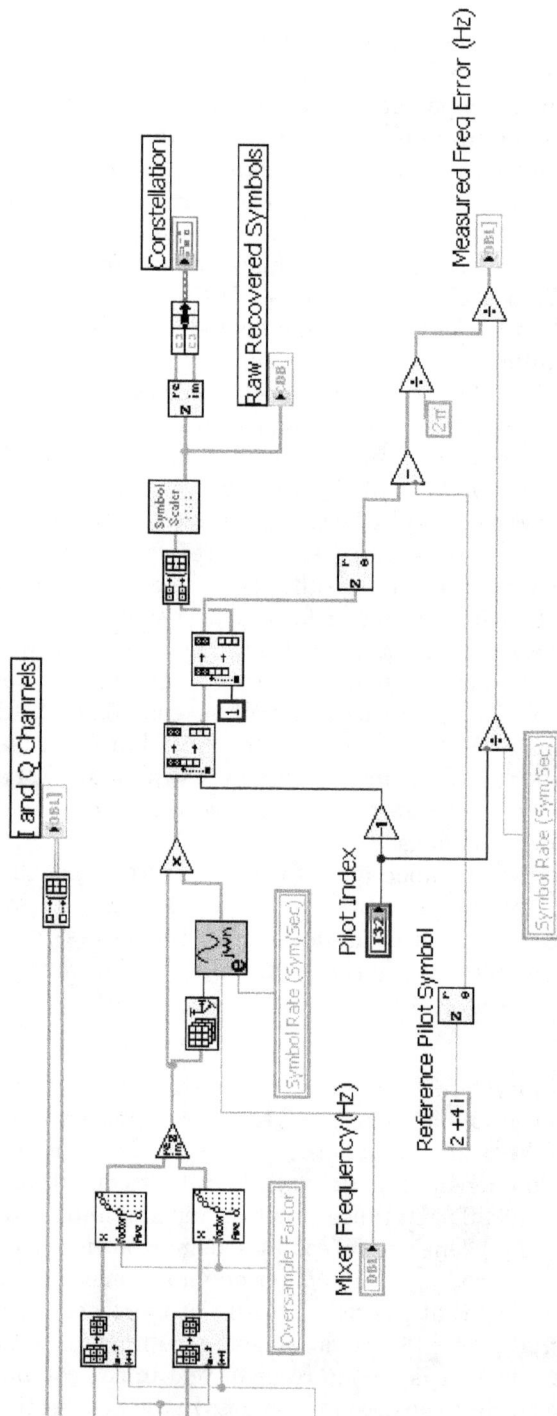

Figure 8.6 A portion of modified Demodulator.vi block diagram.

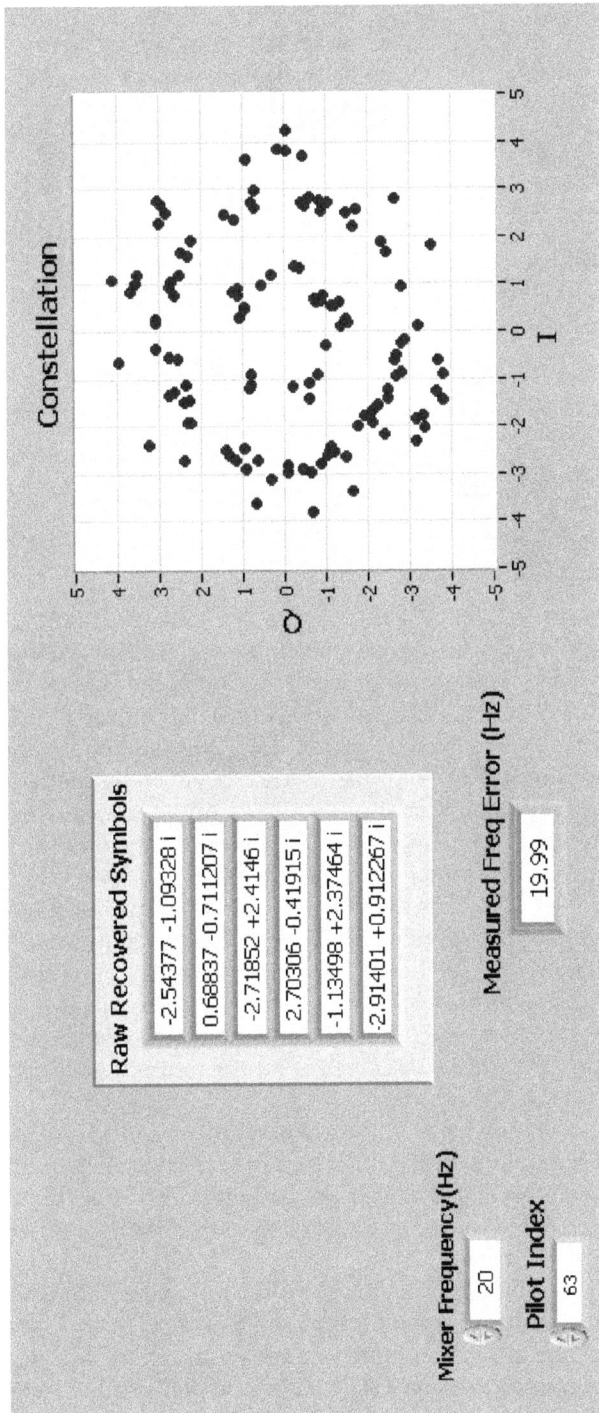

Figure 8.7 Front panel showing demodulated constellation with frequency error.

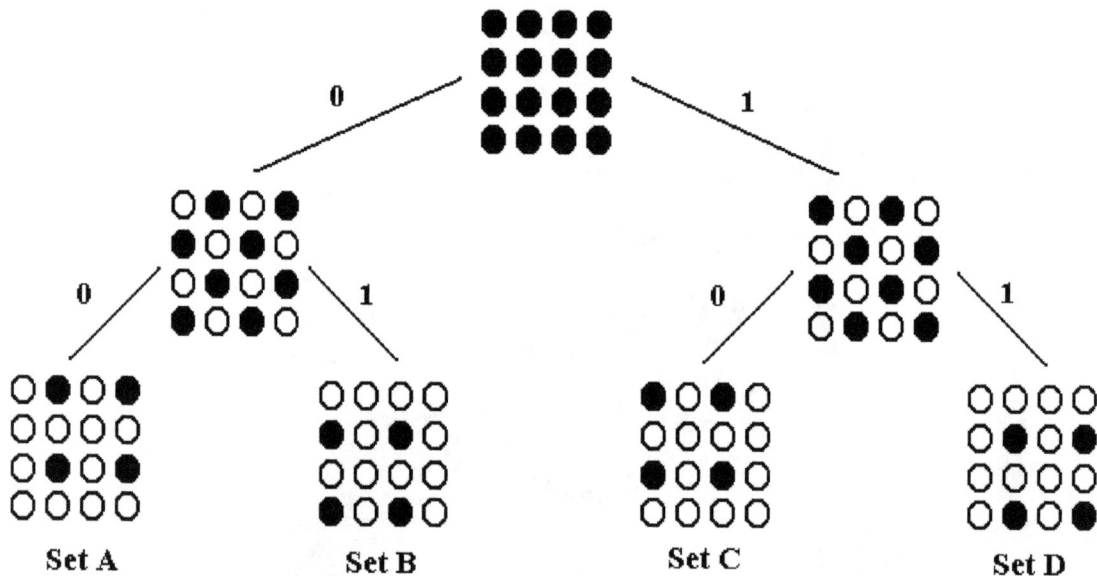

Figure 8.8 Partitioning the 16-QAM constellation.

The initial step is to choose the constellation that we want to transmit. It is important to then properly partition the constellation symbols into sets with the maximum euclidean distance between set members. For a 16-QAM constellation, we will use the mapping from Ref. [2] as shown in Fig. 8.8.

The mapping in Fig. 8.8 tells us that from the current state, an input of 00 moves us to a symbol from set A, while an input of 10 moves us to a symbol from set C. Each of the signal sets comprises four symbols with the maximum euclidean distance between the set members. While Fig. 8.8 shows us the desired constellation partitioning, it does not give us a complete picture of all the state transitions. For that, we need to draw a trellis diagram such as the one in Fig. 8.9. Only a single section of the trellis is shown, the other three sections are similar. Remember that the trellis *completely* defines the behavior of a TCM encoder. The current states are shown on the left and the next states are shown on the right. All of the possible paths from the current states to the following states are the criss-crossing lines in the middle.

Now we are ready to build a state machine in LabVIEW to implement the trellis structure shown in Fig. 8.9. There are several ways to build this state machine with case selectors and conditionals, however, since the trellis is a relatively simple coder, we can eliminate the case selectors and use the virtual instrument (VI) shown in Fig. 8.10.

TrellisStateMachine.vi takes the n-bit sequence from the current state, splits off the oldest k data bits, and appends the current input data bits to form the next state. The signal sets (four in our case), along with the current state and current input bits are used to calculate which symbol to output. More specifically, the current-state bit sequence chooses which of the four signal sets the

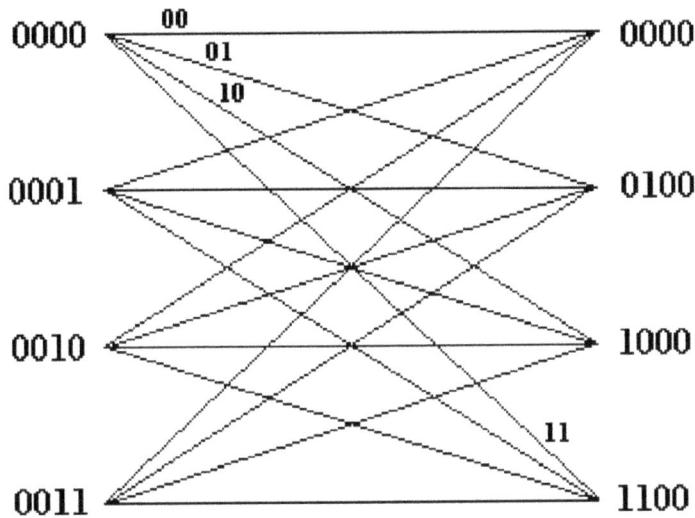

Figure 8.9 Single section of 16 state trellis diagram.

output symbol is in and the current input bit sequence chooses which of the four symbols in the set is chosen as the output symbol. Also you will notice the use of the sub-VI Bin2Dec.vi, which simply converts the input binary sequence to a decimal number (see App. A for more information on this VI).

Now that we are capable of calculating the next state and the current output symbol, we can use TrellisStateMachine.vi to build the complete TCM shown in Fig. 8.11.

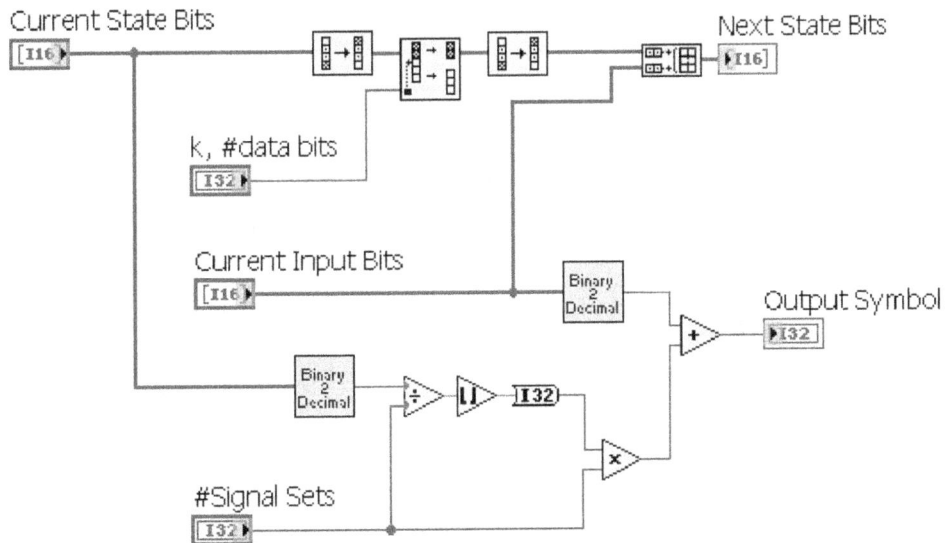

Figure 8.10 TrellisStateMachine.vi block diagram.

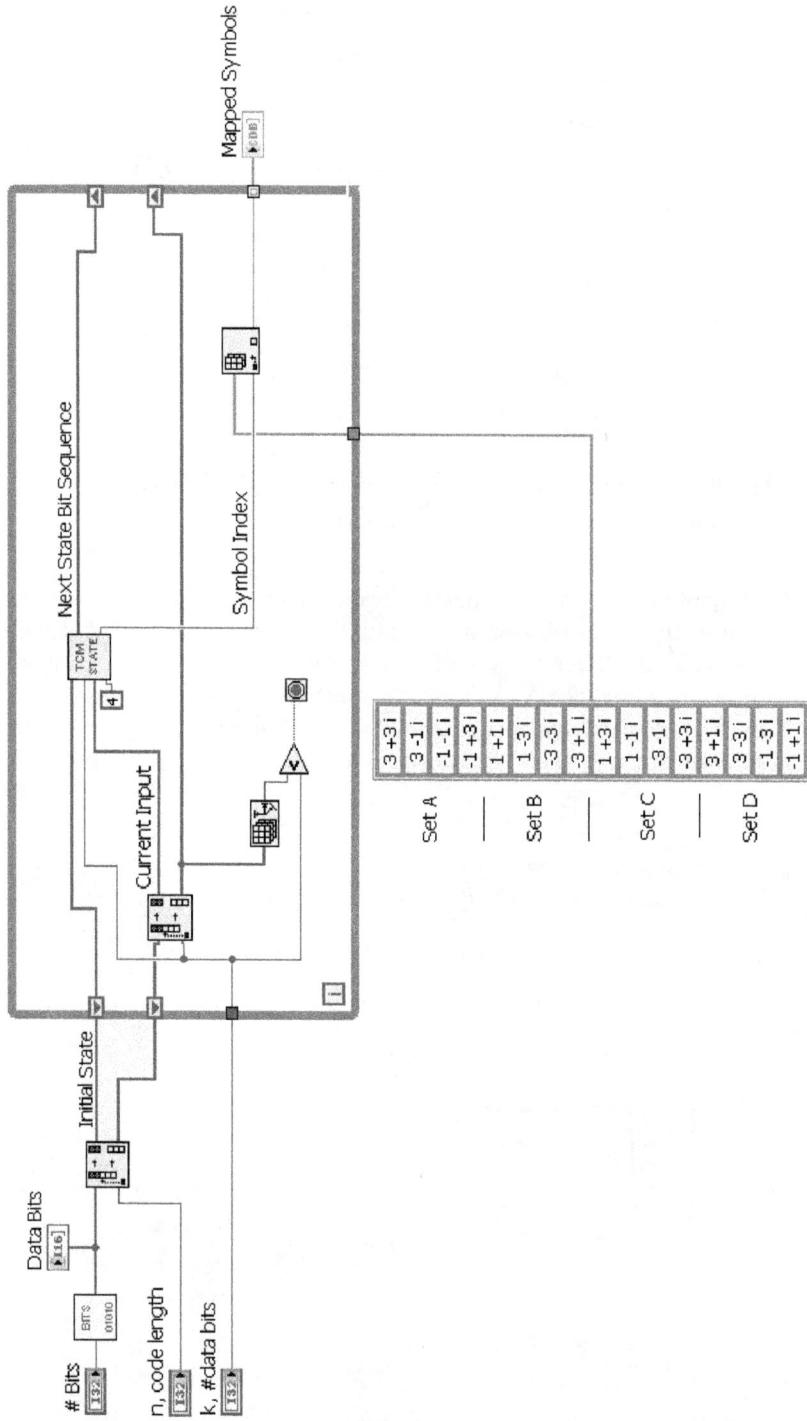

Figure 8.11 TrellisCodedModulator.vi block diagram.

So what does it buy us to use TCM? Hopefully the signal sets that we chose give us increased resistance to bit errors. This means that for the same E_b/N_0 we can expect a lower probability of a bit error. Of course, all this comes at the cost of increased complexity of the receiver.

Speaking of the receiver, we have not looked at what would be involved in decoding these symbols back into their corresponding data bits. Actually decoding a trellis can be quite a lot of work; however, there is an efficient decoding algorithm known as the Viterbi decoder. This is a well-established method for decoding the TCM signal and the next section will partially develop the LabVIEW tools to implement the Viterbi algorithm.

8.2.3   Viterbi decoder

Now comes the difficult part of TCM—decoding the raw symbols. The reason that we get gains in error performance from the trellis is that we are using a memory-based system to make symbol decisions. The trellis gives us somewhat of a roadmap. We can use the known state transitions, along with some kind of distance measure, to give us an overall metric to decide which trellis path is the most likely, thus giving us the best decoded symbol sequence. Let us explore the Viterbi decoder in a step-by-step fashion.

In our 16-QAM signal space, there are four sets {A, B, C, and D} as shown in Fig. 8.11. We can take the first raw recovered signal and choose which symbol is the closest from each of the four signal sets. The block diagram in Fig. 8.12 shows how the euclidean distance between the raw received symbol and each constellation symbol is calculated. The minimum distance is passed as output along with the closest symbol match from each signal set.

Now that we have computed the minimum euclidean distance from our raw symbol to each signal set, we will need to remember these values. Here is where the memory comes in. We are going to store the euclidean distance of the raw symbol from each of the closest symbol sets (four distance measures). And we will keep a running tally of the sum of each path's distance measure. The complicated thing about the trellis is that each of our states has four possible states to transition to. So the number of paths grows by powers of 4 as we step through each state in the trellis. That gets to be an unmanageable number of paths to hang on to and calculate distance metrics for. So what do we do? When we get so far into the trellis (let us say seven transitions) we will choose which path had the lowest overall euclidean distance and therefore that path is the most likely. That way we can throw away all the other paths and focus on just that one. From this discussion you should realize that the Viterbi decoder uses the past information (state memory) to make an informed decision about the best possible match for a transmitted sequence. The user can make their own choice regarding the length of a sequence to match but must keep in mind that the required memory grows quickly with the desired decoder depth.

Figure 8.13 shows the block diagram of a partial Viterbi decoder. It is incomplete in the sense that it is only a starting place for the readers to begin building

Raw Received Symbol

CSG ▶

Set A

| 3 +3 i |
| 3 -1 i |
| -1 -1 i |
| -1 +3 i |

Set A Distance

▶SGL

Set A Element

▶CSG

Set B

| 1 +1 i |
| 1 -3 i |
| -3 -3 i |
| -3 +1 i |

Set B Distance

▶SGL

Set B Element

▶CSG

Set C

| 1 +3 i |
| 1 -1 i |
| -3 -1 i |
| -3 +3 i |

Set C Distance

▶SGL

Set C Element

▶CSG

Set D

| 3 +1 i |
| 3 -3 i |
| -1 -3 i |
| -1 +1 i |

Set D Distance

▶SGL

Set D Element

▶CSG

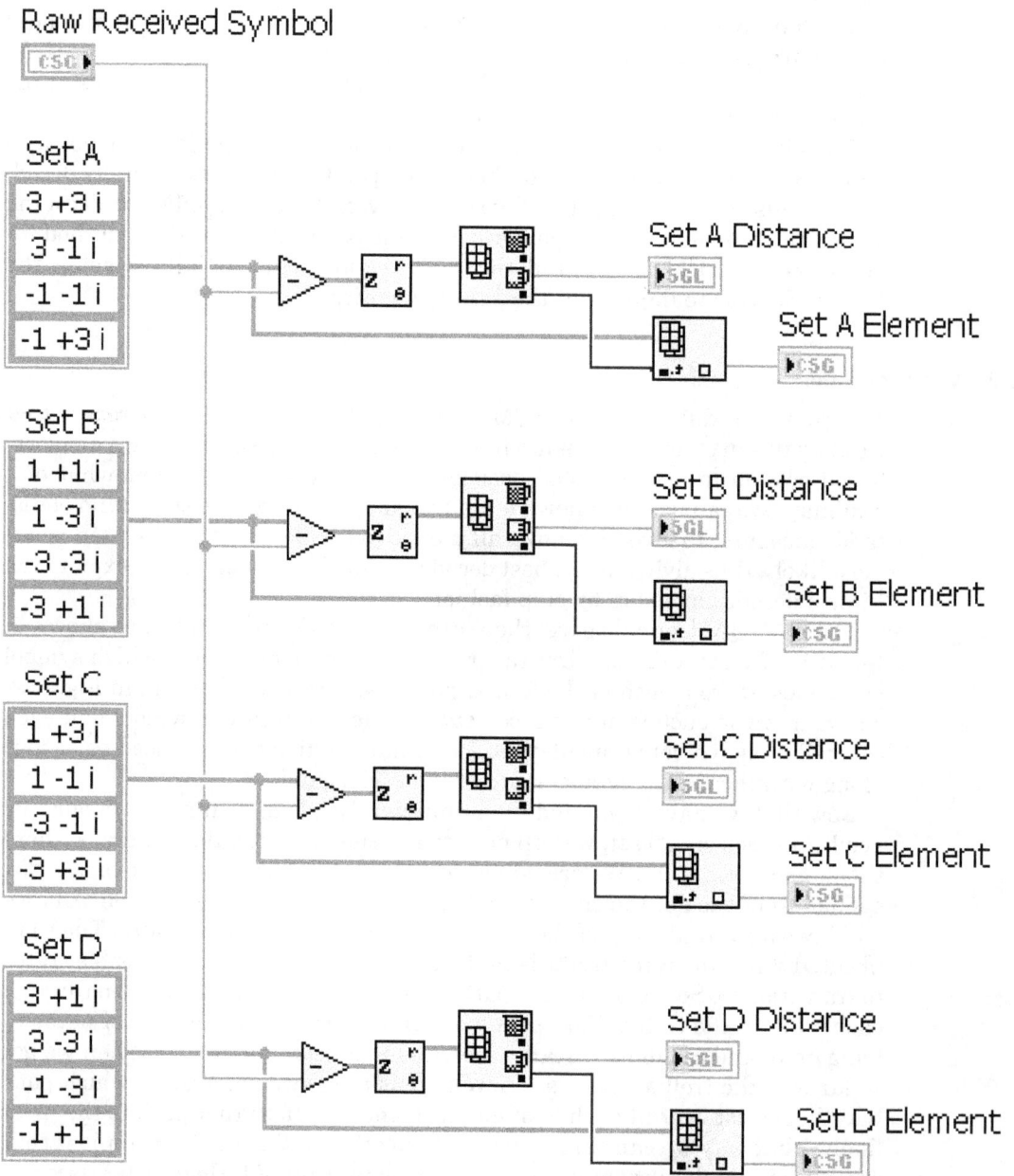

Figure 8.12 Viterbi_distance.vi block diagram.

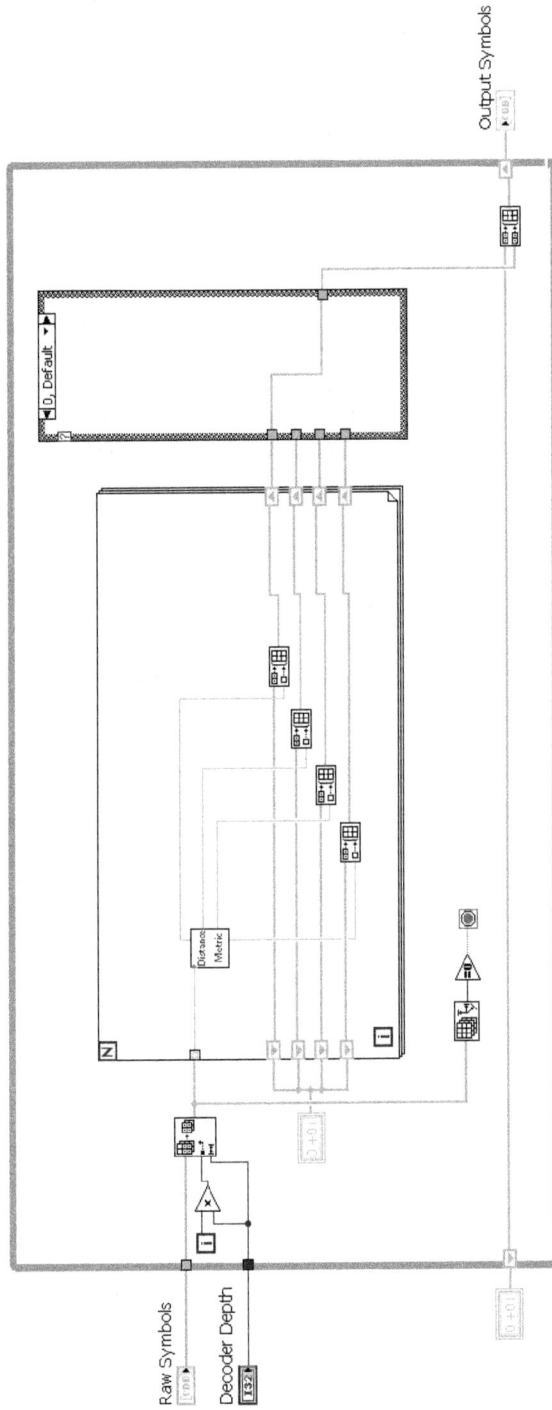

Figure 8.13 Partial Viterbi decoder block diagram.

their own decoder. The VI Viterbi_distance.vi is placed inside a loop that runs over the decoder depth. The euclidean distance measures are stored using shift registers. In order for this VI to be complete, the closest matched symbols from each set will have to be stored and the next symbol can only be chosen from the trellis. It will be up to the reader to complete the Viterbi decoder.

Summary

This chapter has developed some tools for measuring the performance of a digital communication system. Specifically, we saw VIs calculate the number of bit errors and the bit error rate, as well as the error vector magnitude. Together these tools should give us a very good indication of the performance of any digital communication system. On top of that, some methods for improving system performance such as channel estimation and coding were presented. Also TCM and the Viterbi decoder were examined and LabVIEW examples were developed for implementing those concepts.

References

1. Proakis, J. G., *Digital Communications,* 4th ed., McGraw-Hill, New York, 2001.
2. Agilent Technologies PN 89400-14, "Using Error Vector Magnitude Measurements to Analyze and Troubleshoot Vector-Modulated Signals."
3. Sklar, B., *Digital Communications,* 2d ed., Prentice-Hall, Upper Saddle River, NJ, 2001.

Optimizing LabVIEW
Signal Processing

In general, most of the processing of large signals done in LabVIEW is fairly slow. There is overhead from the personal computer's (PC's) operating system, from the LabVIEW environment, from updating the front panel display, and then all of the filtering and data manipulation that goes along with demodulating your signal. We have already seen that one way to make *huge* savings on the processing end is to lower the sample rate of the incoming signal. Of course, that is not always a practical solution, so in this chapter we cover some generic tips for streamlining LabVIEW coding as well as some neat tricks for doing fast signal processing.

9.1 General LabVIEW Coding Guidelines

As with most programming environments, there are some simple things you can do to maximize the efficiency of your code. This list is by no means exhaustive, but these are a few things that I have used to streamline my own programs and will save you time with very little effort.

1. *Watch what you put inside a loop:* Repeated computations should always go outside a loop: This should be a standard practice for any programming language including LabVIEW. Do not be tempted to put a bunch of operations (or sub-VI calls) inside a loop if they do not depend on the loop or need to be updated on each pass. Also you have to be careful where you place controls. Look at Fig. 9.1 that shows a simple convolution block diagram.

Both input controls are placed inside the loop in Fig. 9.1. This means that on each loop iteration, the inputs will be scanned. Since the filter coefficients will

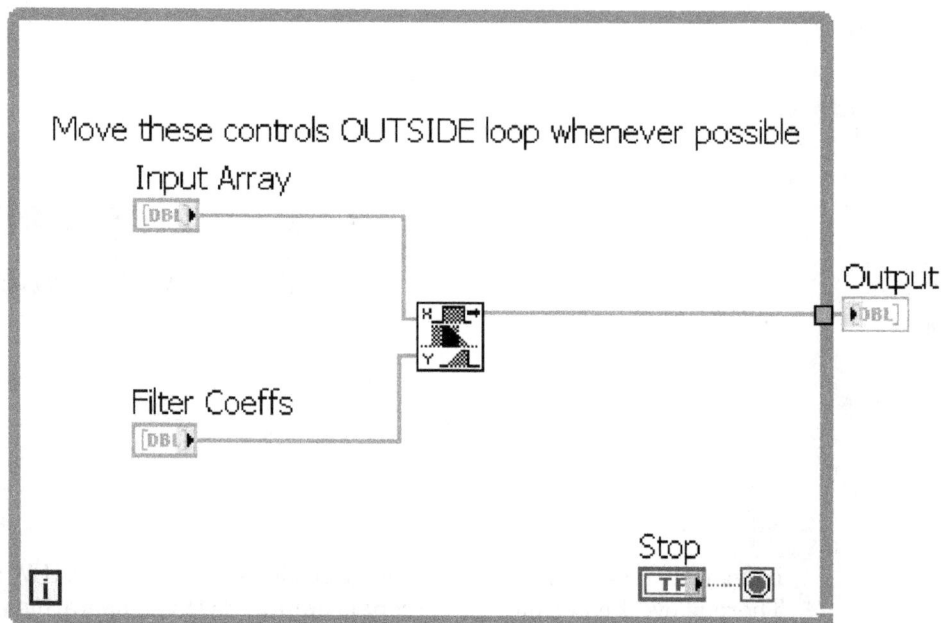

Figure 9.1 Simple convolution loop block diagram.

probably not change, this control really should be placed outside the loop. If the input is expected to change on each loop iteration, then it should stay inside the loop.

2. *Precompute as much as possible:* This applies to filter coefficients, or complex mixer values, or any set of values that will probably not change during runtime. There is certainly no point in recomputing a set of filter coefficients or the fast Fourier transform (FFT) of a fixed data sequence.

3. *Avoid the use of global and local variables:* Use sequence locals instead of local variables and in general try to avoid global variables in speed critical applications. This one gives me a lot of trouble because I really like the readability that you get from using global and local variables. They are self-labeling, you can instantly tell if they are read or write and they make for very clean looking VIs without wires running all over—but, they are terribly slow. It is much more efficient to use sequence locals instead. I have been told that every instance of a LabVIEW global variable is stored separately in the memory. If you use large arrays, you definitely do not want to reference them with global or local variables.

4. *Minimize the number of display updates and the amount of data displayed to the front panel:* The LabVIEW graphics are great and it sure is nice to use

them, but displaying data on the screen is *slow*. Unless the plot is critical to the application, I would skip it. Just keep the display data to a minimum and you will notice huge differences in execution speed.

5. *Use built-in virtual instruments (VIs) whenever possible:* If you closely examine most of the built-in LabVIEW functions, you will notice that the underlying code is not another VI, but instead it is a call to a C routine. Those calls to a C routine are usually *much* faster than if you were to build an equivalent VI in LabVIEW. Sure it makes perfect sense, but you may be tempted to sometimes throw together your own VI to perform a certain function. It is a good idea to check first to see if LabVIEW already has one there.

6. *Build executable applications:* LabVIEW has the capability to build a VI into an executable application. This is an add-on piece of software known as the application builder. What this does is allow anyone to *run* the application without having a copy of LabVIEW. They do have to download the LabVIEW run-time engine. While this method is not guaranteed to speed up your application, it *should* reduce the required memory consumption because the full LabVIEW suite is not loaded and thereby give your PC some breathing room for slightly faster execution. Now of course if your PC is very powerful with plenty of memory you may not notice much improvement here. But it is certainly worth a try and the bonus is that no one can view or modify your block diagrams.

9.2 Signal Processing Tips

Other than using the previously mentioned tips for coding speed, there are a few ways to increase the speed of some common digital signal processing (DSP) calculations. The two things you may find yourself doing more than anything else are: (1) filtering large input arrays and (2) computing the FFT. Following are some quick tips for performing these operations efficiently in LabVIEW.

9.2.1 Linear convolution with the FFT

You may recall that convolution in the time domain becomes multiplication in the frequency domain. What this means is that we can perform filtering in the frequency domain by multiplying the discrete Fourier transform (DFT) of our filter coefficients by the DFT of the signal. Now assuming the input is a fairly long sequence and we use the FFT to compute the DFTs, this can be a significant saving. Before we look at performing this in LabVIEW, let us take a moment to review the restrictions on using the DFT for linear convolution. Linear convolution takes inputs of lengths L and N and produces an output of length $L + N - 1$ [1]. To use the FFT to get the same output, the length *must* be at least $L + N - 1$. Simple enough—just pad the inputs to the next power of 2 that is *at least* that length.

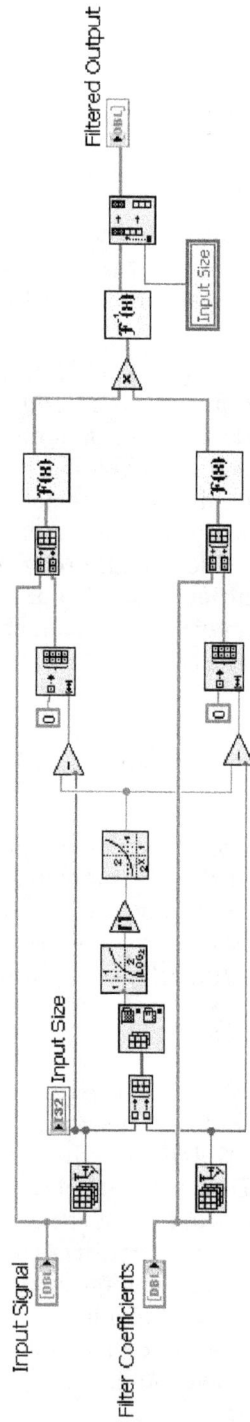

Figure 9.2 Conv_with_FFT.vi block diagram.

The block diagram in Fig. 9.2 shows that the input signal and the filter coefficients are padded up to the next power of 2 size. The chosen power of 2 is based on the larger of either the filter array or the input array. The FFT outputs are then multiplied and the inverse FFT is computed. For even greater savings, the FFT of the filter coefficients needs to be performed only once, while the input data FFT is performed on each sample interval.

9.2.2 Fast real FFT

Using the FFT is pretty fast, but it could be faster by packing the real and imaginary parts of a complex FFT input array with the even and odd (respectively) components of a real signal. When this is done, an N-point real FFT can be replaced with an $N/2$-point complex FFT [2]. The first step here is to separate the input sequence into its even and odd indexed components. Then reassemble those two component sequences into a single complex sequence. Now we can perform an $N/2$-point complex FFT. However, there is a bit of juggling to do with the output—namely, you now have an $N/2$-point FFT that is somehow supposed to represent the FFT of an N-point data sequence. Using the properties of the Fourier transform, you then generate the other half of the FFT [2]. Figure 9.3 shows the block diagram of FastRealFFT.vi.

FastRealFFT.vi uses the equations from Ref. [2] for separating this complex FFT back into the components of our real signal's spectrum. There is an intermediate step involved where values must be stored and then operated on to separate the complex FFT back into the components of the real signal's FFT. This example has not been optimized, but it shows how the complex FFT can be used to generate the FFT of a real signal.

9.3 More LabVIEW DSP Applications

The tools and functions that we have seen so far are just a small example what LabVIEW can do. As we will see in this section, LabVIEW has palettes full of mathematic functions. Here we have access to functions for curve fitting, calculus (including solving differential equations), root solving, approximating functions, and a host of special functions like the Bessel function and complementary error function. The location of these functions is shown in Fig. 9.4.

Some of these functions will be of more use to us than others, in particular the zero solving functions and the linear algebra tools. The location of each of these subpalettes is shown in Figs. 9.5 and 9.6. The next two sections make use of some of these functions.

9.3.1 Roots of difference equations

LabVIEW has a nice built-in function to find the complex roots of a polynomial. This function can be extremely useful when we want to examine the roots of a

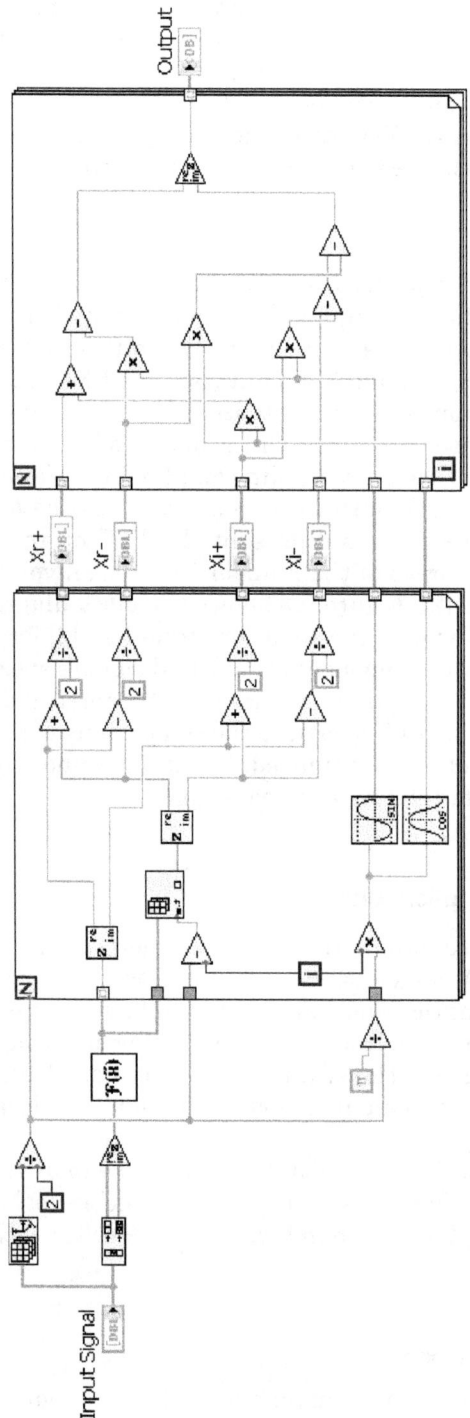

Figure 9.3 FastRealFFT.vi block diagram.

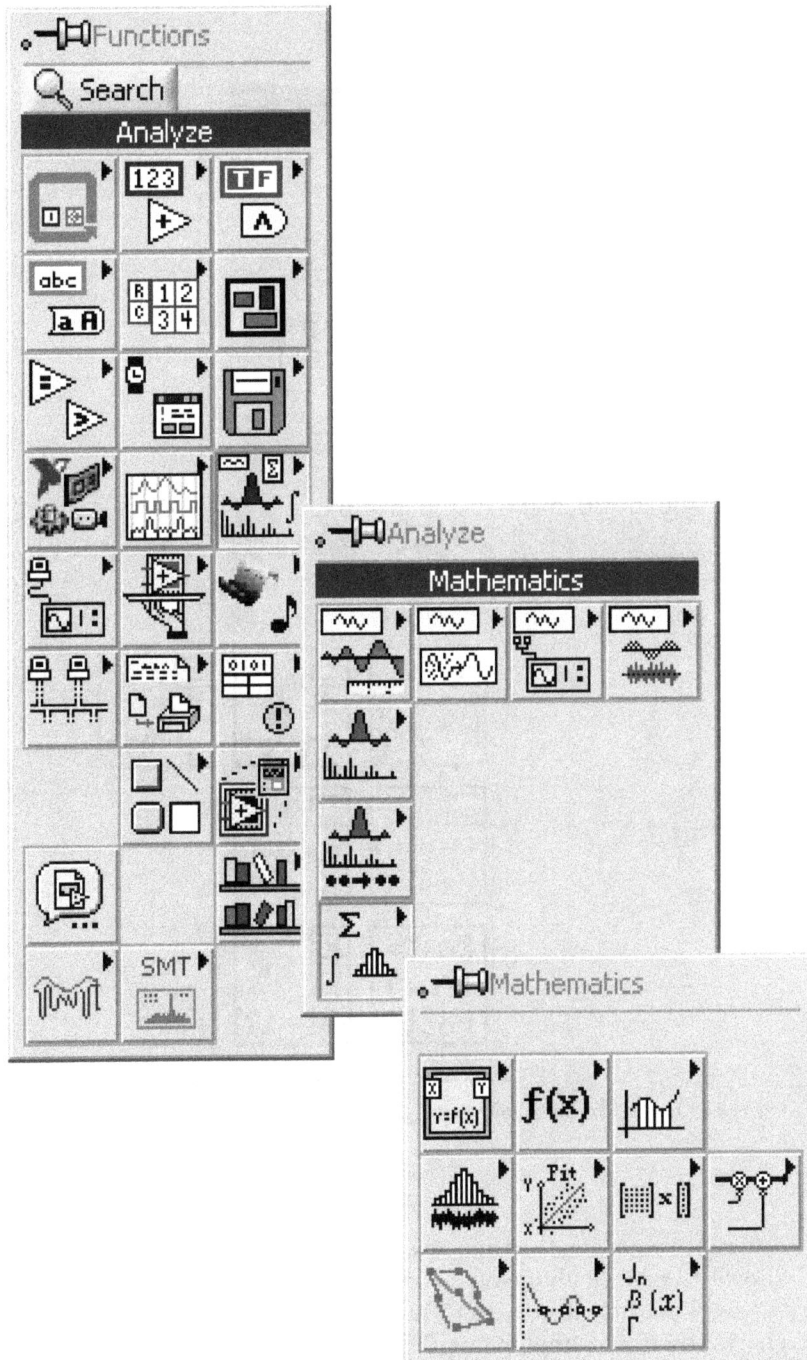

Figure 9.4 Location of LabVIEW Mathematics palette.

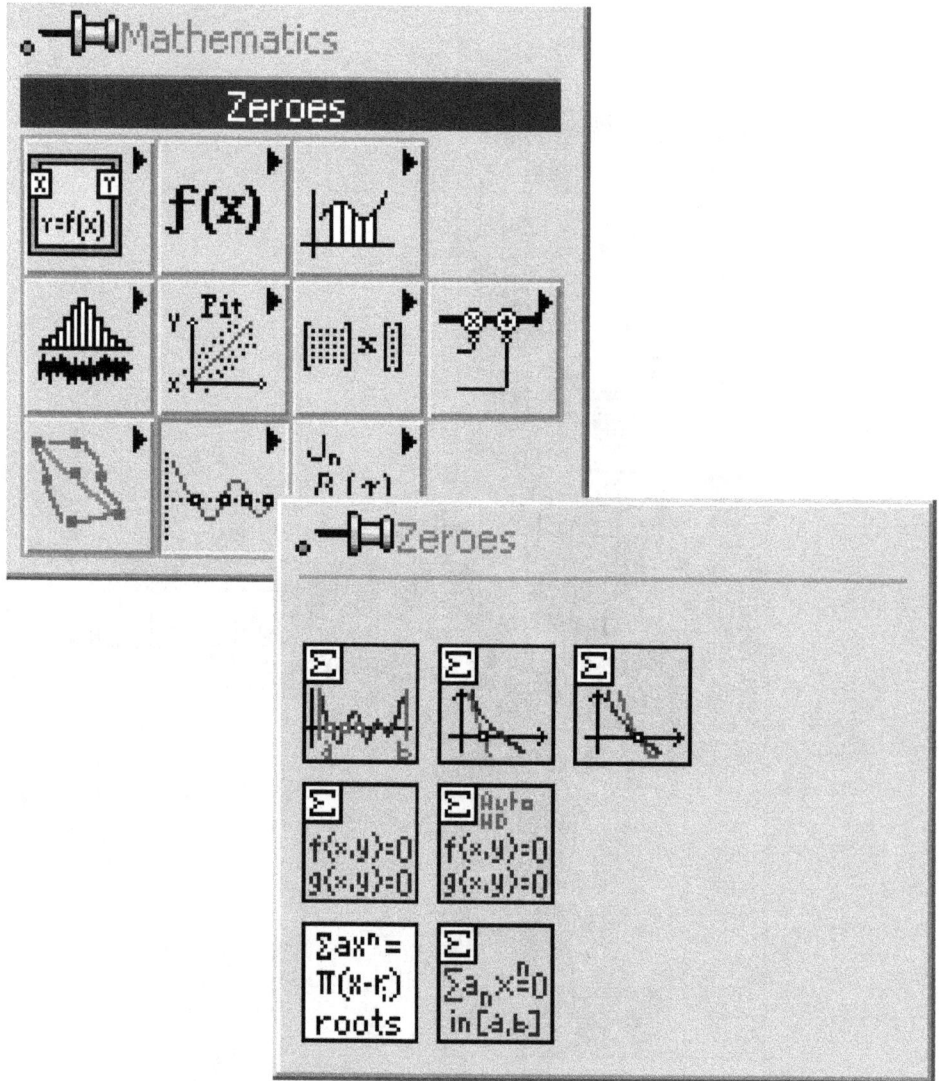

Figure 9.5 Location of LabVIEW zeroes functions.

difference equation that may describe the transfer function of a filter. Figure 9.7 shows the block diagram of PoleZero.vi, which uses Complex Polynomial Roots.vi in order to solve a difference equation entered by the user. Figure 9.8 shows the front panel of our PoleZero VI. Here we can see that the difference equation coefficients *must* be entered in descending powers of z. The graph shows the pole and zero locations for the given difference equation in relation to the unit circle.

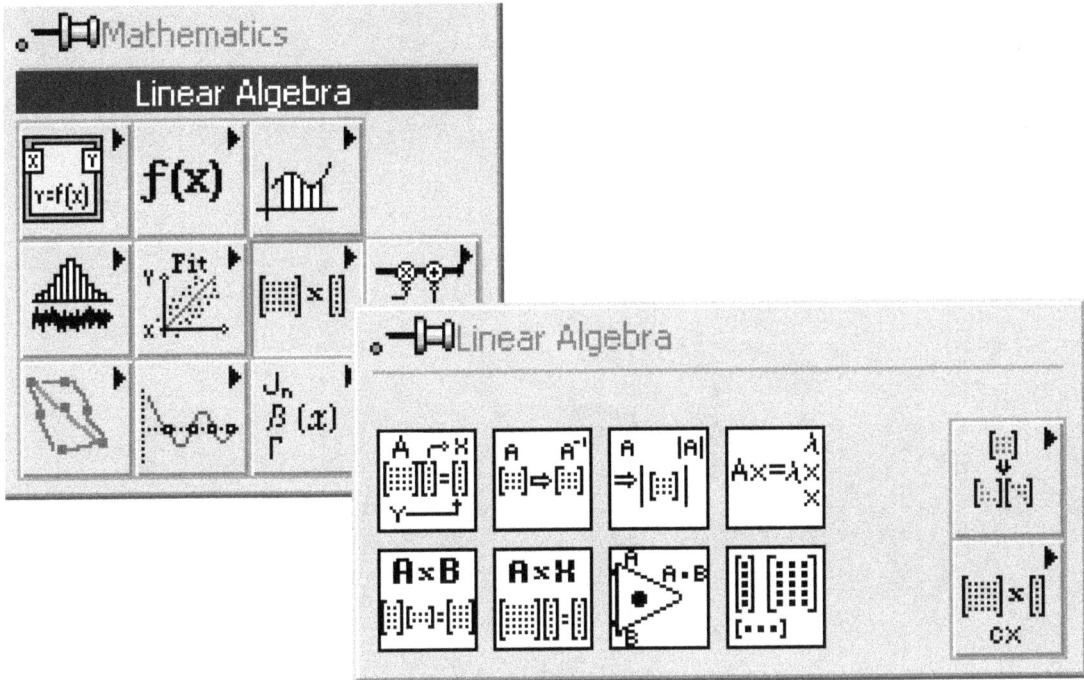

Figure 9.6 Location of LabVIEW linear algebra functions.

9.3.2 Linear predictive speech coder

All digital communication systems employ some form of speech coding to reduce the required bit rate for a voice transmission. Here we will take a look at one of the more common speech coding algorithms, linear predictive coding (LPC). The LPC vocoder is based on the premise that a short time segment of speech can be modeled as the output of a linear filter excited by periodically spaced pulses [3]. There are many more subtleties to actually implementing this kind of speech coding, so if you really want to understand the LPC vocoder, there is an entire chapter in Ref. [4] devoted to the topic. This section presents a LabVIEW implementation of only the central portion of the LPC vocoder, the coefficient generator.

In order to model a short voice segment as the output of a linear filter, we must find the filter coefficients that most closely match the spectrum of that voice segment. So this is just a problem in error minimization that can be expressed by the following equation [4].

$$\sum_{k=1}^{p} \alpha_k R_n(|i-k|) = R_n(i) \qquad 1 \le i \le p \qquad (9.1)$$

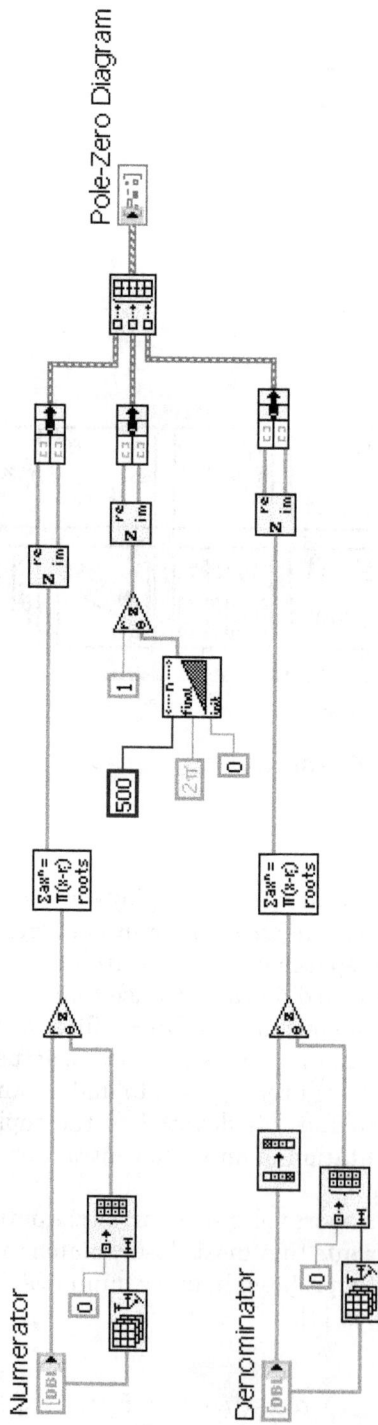

Figure 9.7 Block diagram of PoleZero.vi.

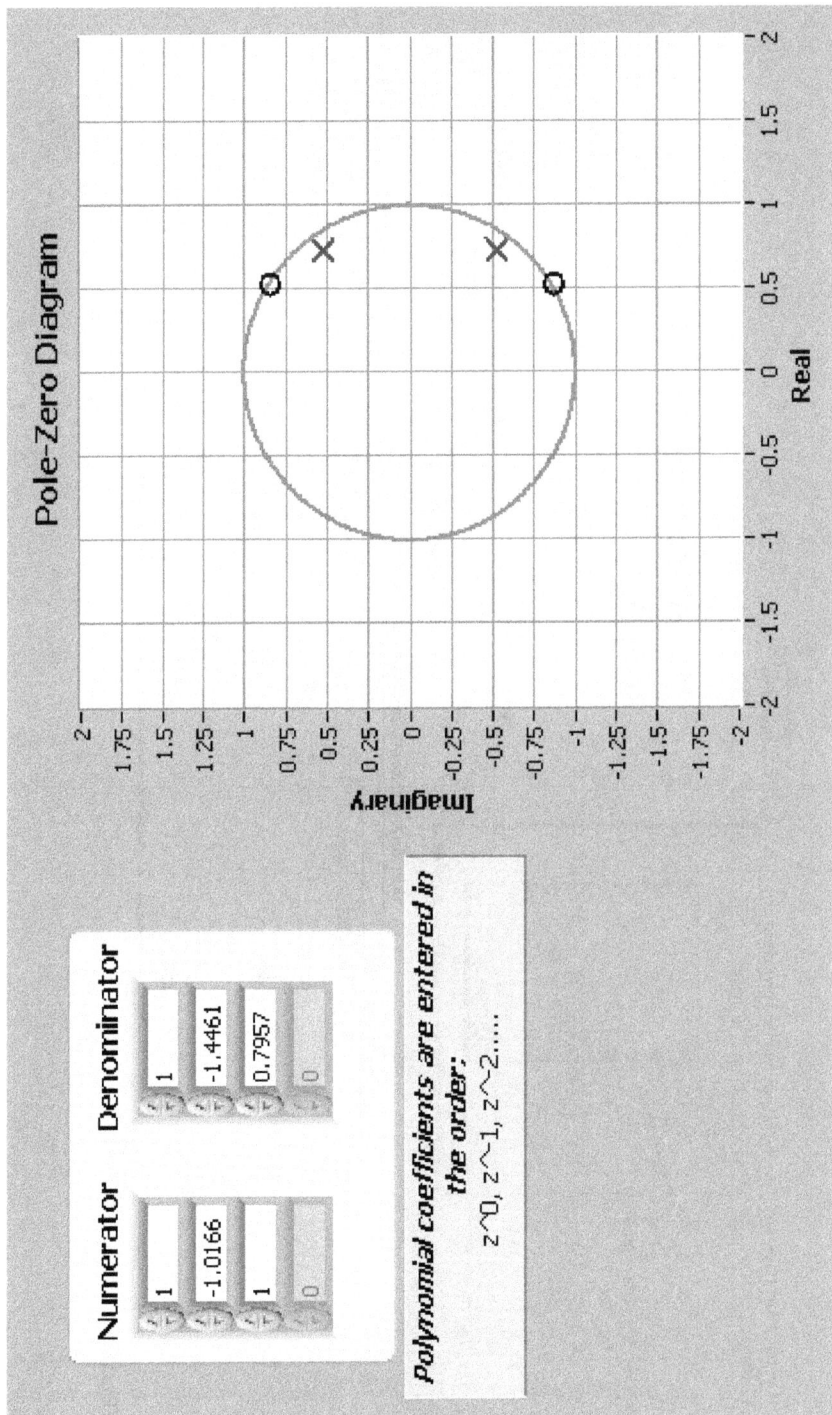

Figure 9.8 Front panel of PoleZero.vi.

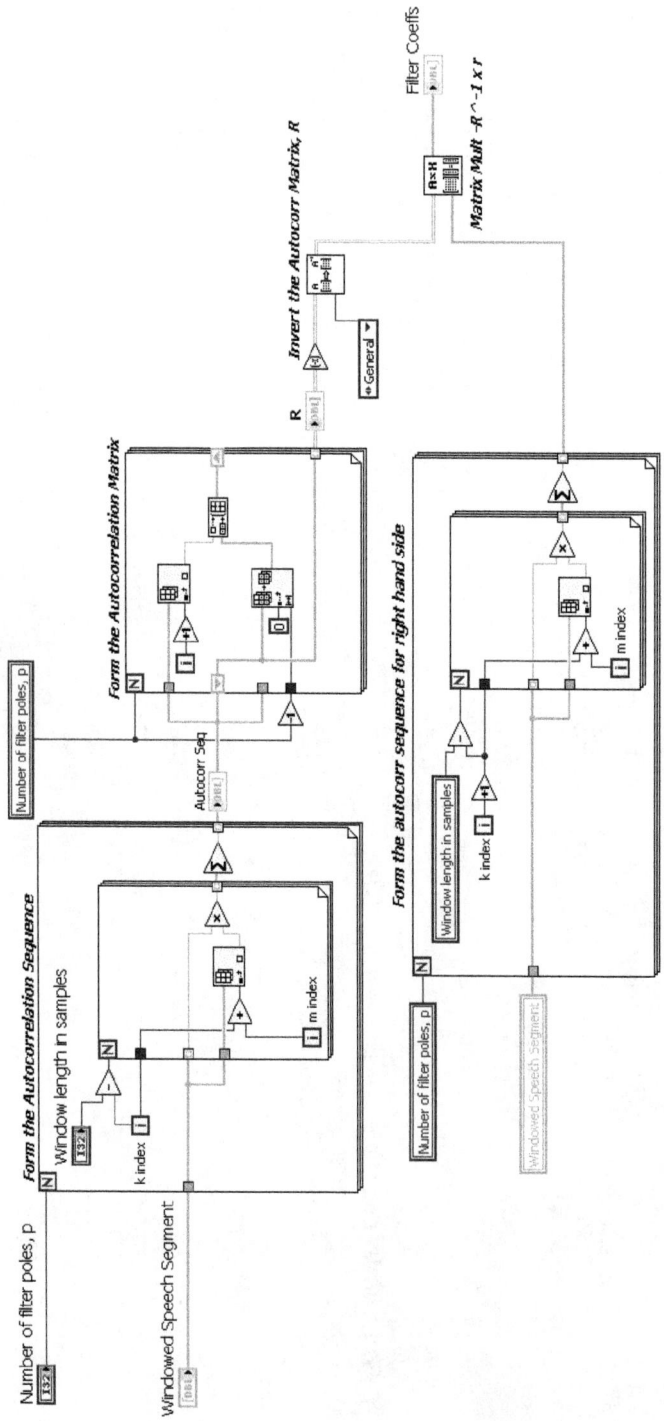

Figure 9.9 LPCcoeffs.vi block diagram.

where p is the number of poles in the filter model and R_n is given by [4]

$$R_n(k) = \sum_{m=0}^{N-1-k} s_n(m)s_n(m+k) \qquad (9.2)$$

and $s_n(m)$ is the windowed speech signal.

There are some very efficient methods for solving the set of linear equations, one of which is the Levinson–Durbin recursion. However, since LabVIEW has some built-in matrix manipulation tools, we can just invert the autocorrelation matrix as shown in the block diagram of LPCcoeffs.vi in Fig. 9.9.

The first step in LPCcoeffs.vi is to compute the short time autocorrelation function $R_n(k)$ from Eq. (9.2). Once we have the autocorrelation sequence, we can use those values to form the autocorrelation matrix (which is a Toeplitz matrix). From here, the LabVIEW matrix inversion function is called to invert the (sign negated) autocorrelation matrix. At the same time, another autocorrelation sequence is formed (this time from lag 1 to lag p) and this sequence is matrix multiplied by the inverted autocorrelation matrix. The result of this multiplication is the sequence of p filter coefficients.

The LPCcoeffs.vi can now be dropped into another VI that will sample the audio from the soundcard (perhaps use sndcard.vi shown in Chap. 2), perform the windowing, and then use the filter coefficients output from this VI to regenerate the voice frame (or convert them into reflection coefficients for a lattice implementation). Of course there is more to the LPC vocoder such as voiced/ unvoiced decisions, filter gain, and some kind of pitch detection algorithm, but LPCcoeffs.vi should get you started.

Summary

This chapter has presented some basic LabVIEW coding techniques for increasing the speed of your routines. Also shown were a couple of computational tricks for a few of the most common DSP computations. Finally, we ended up with some discussion on the mathematical abilities of LabVIEW and saw two DSP applications of LabVIEW outside of the digital communication's focus in the rest of this book. There are many more functions built into LabVIEW than could be covered by this book; however, it is my hope that the topics included here will start the reader down the path of using the full power of LabVIEW.

References

1. Oppenheim, A. V., R. W. Schafer, and J. R. Buck, *Discrete-Time Signal Processing,* 2d ed., Prentice-Hall, Upper Saddle River, NJ, 1998.
2. Lyons, R. G., *Understanding Digital Signal Processing,* Prentice-Hall, Upper Saddle River, NJ, 2001.
3. Ingle, V. K., and J. G. Proakis, *Digital Signal Processing using Matlab,* Brooks/Cole, Pacific Grove, CA, 2000.
4. Rabiner, L. R., and R. W. Schafer, *Digital Processing of Speech Signals,* Prentice-Hall, Upper Saddle River, NJ, 1978.

VI Reference

1. AdvFFT.vi
10. ComplexMixer.vi
19. Halfband.vi
28. PoleZero.vi
37. SincFcn.vi

38. SoundCardCapture.vi

2. ArraySwap.vi
11. Conv_with_FFT.vi
20. IIR.vi
29. PolyphaseInterpolator.vi
39. SymbolDecoder.vi

3. AWGN.vi
12. Demodulator.vi
21. Init5660.vi
30. PulseFreq.vi
40. SymbolMapper.vi

4. BasicFcns.vi
13. Downsample.vi
22. KaiserFIR.vi
31. PXICapture.vi
41. SymbolRateAxis.vi

5. BERTest.vi
14. EVM.vi
23. LPCcoeffs.vi
32. Rayleigh.vi
42. TrellisCodedModulator.vi

6. Bin2Dec.vi
15. FadeSimulation.vi
24. MatchedFilterDetection.vi
33. ScaleSymbols.vi
43. TrellisStateMachine.vi

7. BoundaryDecoder.vi
16. FastRealFFT.vi
25. Modulator.vi
34. Set0dB.vi
44. Upsample.vi

8. Channel.vi
17. FreqAxis.vi
26. MPR.vi
35. SimpleFFT.vi
45. Viterbi_distance.vi

9. Chirp.vi
18. GenerateBits.vi
27. NyquistPulse.vi
36. SimpleNoisySystem.vi
46. Window.vi

1. **AdvFFT.vi.**   An advanced FFT utility. This VI has the ability to force the use of the FFT algorithm by padding the input length to a power of 2. The input can either be real or complex and the FFT output is shifted over by half the sample rate and plotted versus frequency.

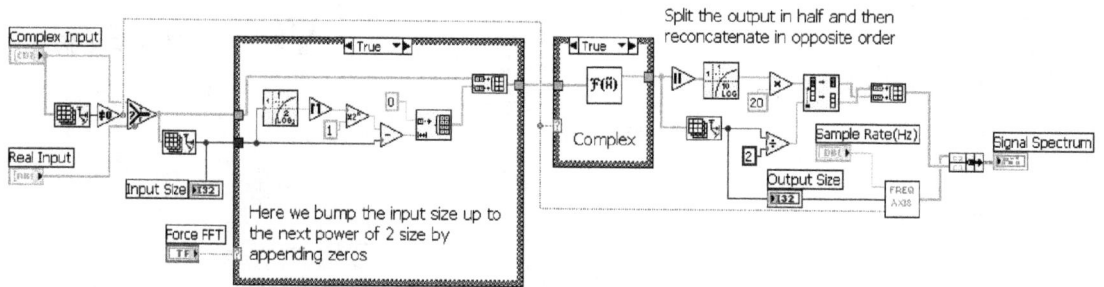

2. **ArraySwap.vi.**   A simple utility to switch the positive and negative halves of the FFT output (or any array for that matter).

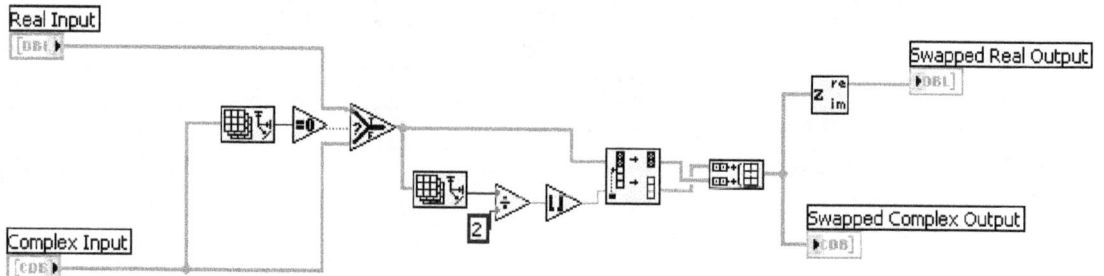

3. AWGN.vi. AWGN generates white gaussian noise with a specified noise power. The generated noise can be either real or complex.

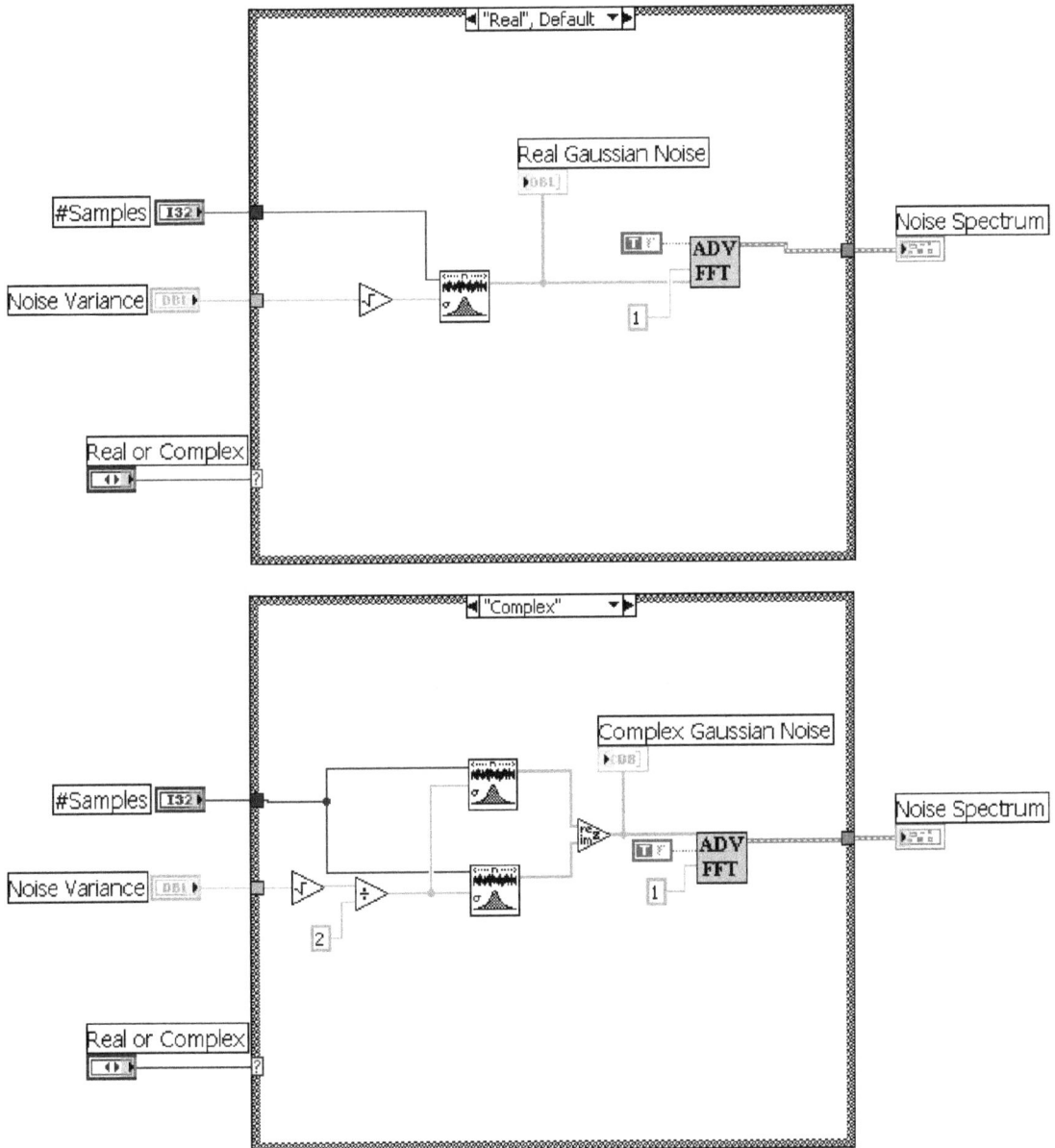

4. BasicFcns.vi. Forms the impulse, ramp, and step functions with a specified number of samples and delay. The amplitude is fixed to 1.

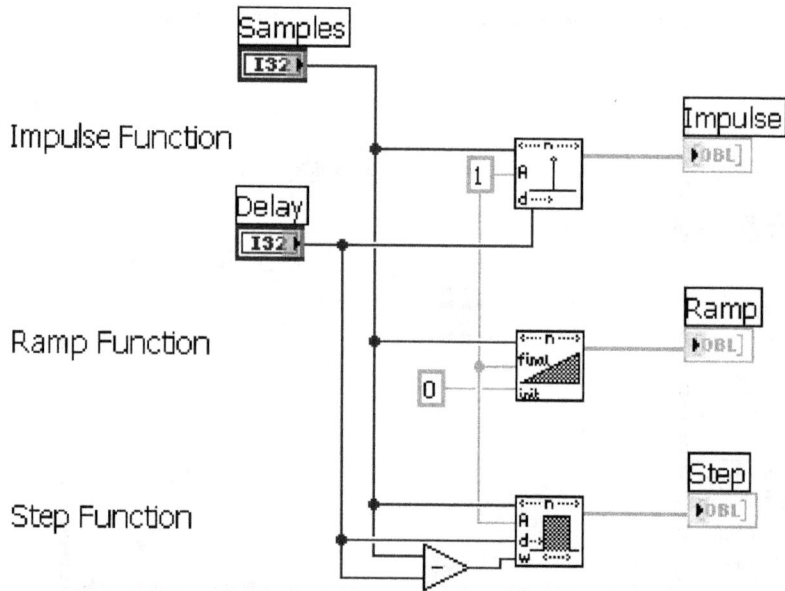

5. BERTest.vi. This VI calculates the number of bit errors as well as the bit error rate of our 16-QAM waveform. The known bit sequence is output from the modulator and compared to the recovered bit sequence after the waveform is corrupted by Channel.vi.

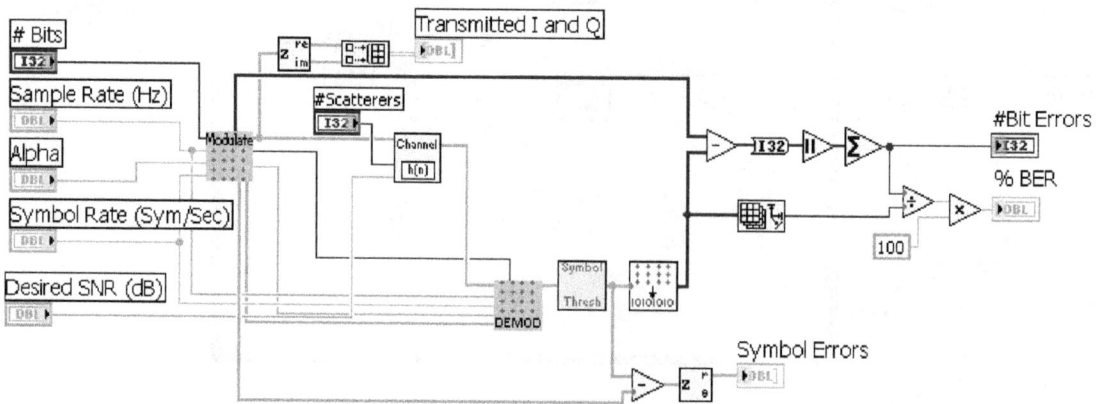

6. Bin2Dec.vi. Converts a binary input array to a decimal number. The highest index is the most significant bit and index 0 is the least significant bit.

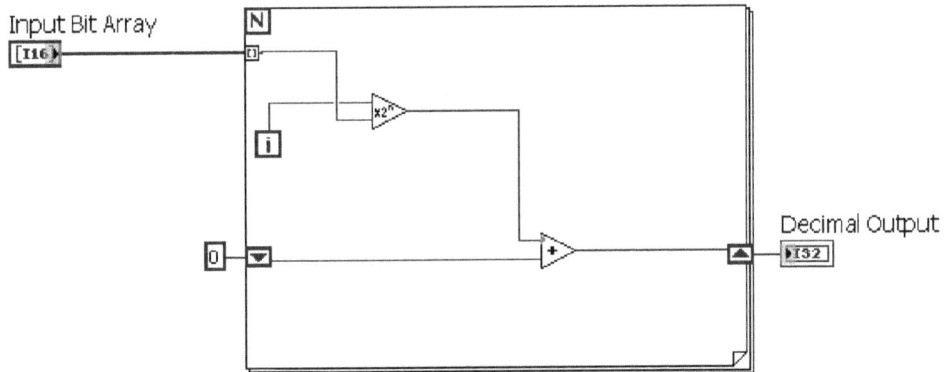

7. BoundaryDecoder.vi. This VI chooses the closest constellation symbol match for the recovered symbols based on boundary decisions. For this 16-QAM constellation, the regions are 0 to ± 1 and $\geq \pm 2$.

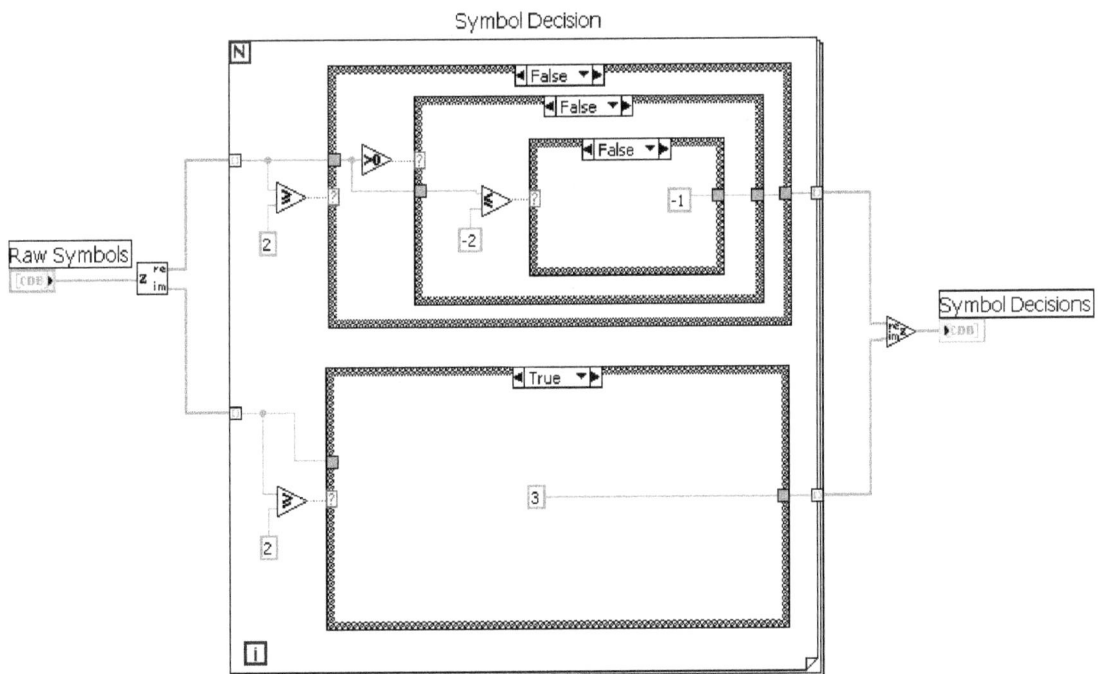

8. Channel.vi. This VI simulates a fading channel by forming a Rayleigh faded envelope and applying the faded profile to the modulated waveform.

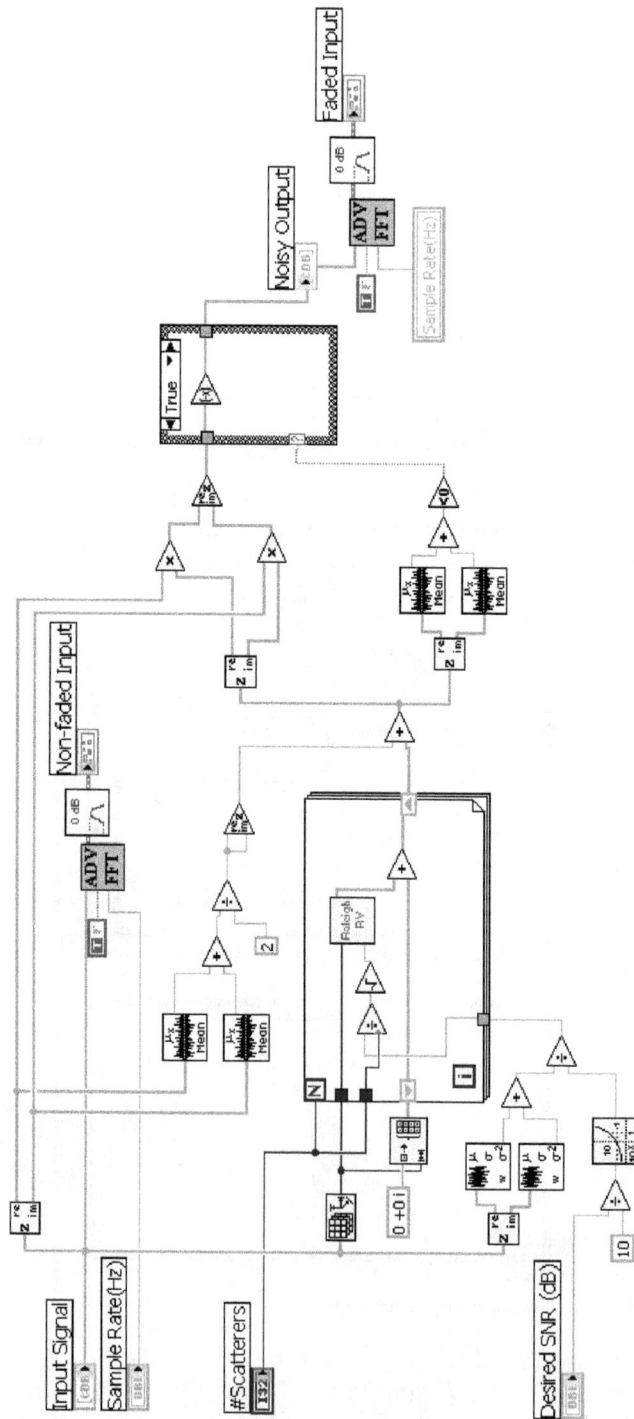

9. Chirp.vi. Computes a linear chirp sequence where the instantaneous frequency of the waveform linearly increases with slope k.

Linear Chirp

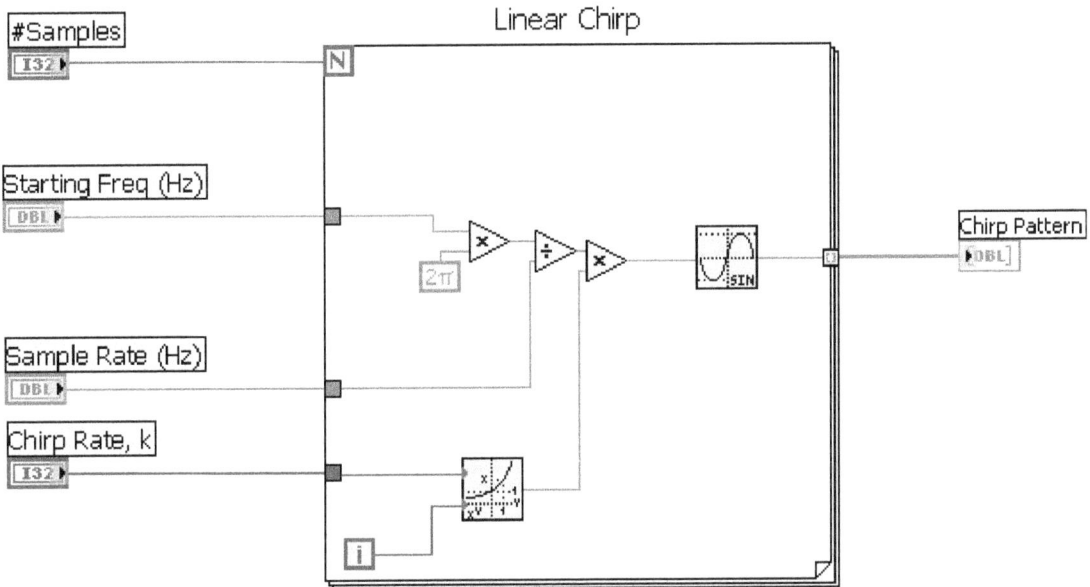

10. ComplexMixer.vi. Generates a complex sinusoidal mixer of a given length and frequency for the specified sample rate.

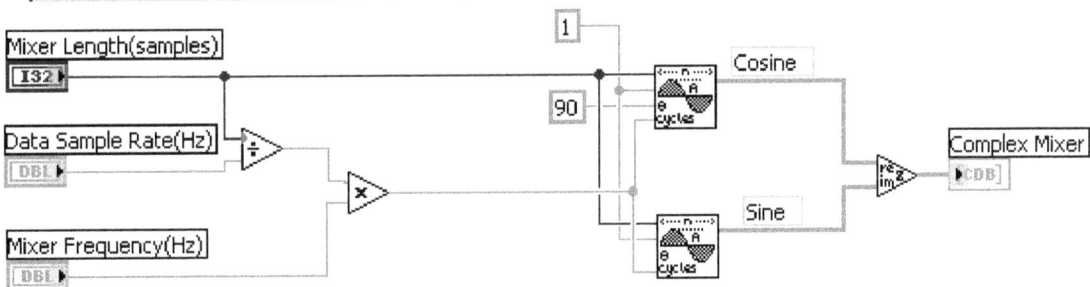

11. Conv_with_FFT.vi. Performs linear convolution with FFT. The input signal and filter coefficients are zero-padded up to the next highest power of 2 length. The FFTs of the extended sequences are multiplied and the inverse FFT is calculated. The output is then trimmed to remove excess zeros.

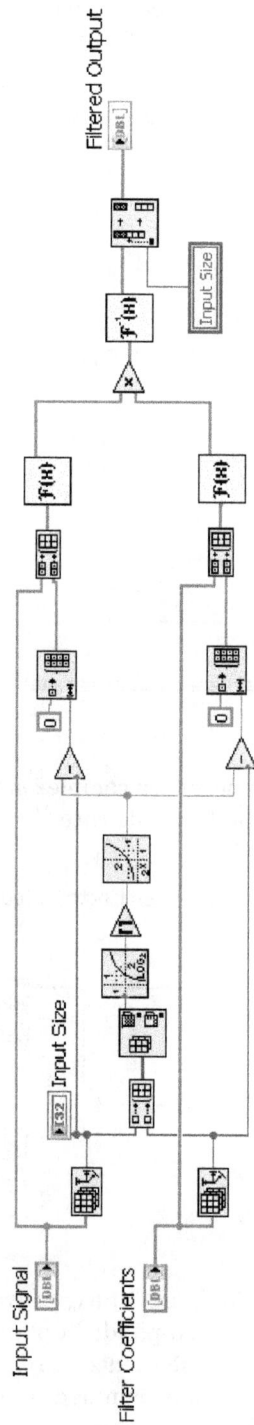

Filtered Output

Input Size

Input Signal

Filter Coefficients

Input Size

12. Demodulator.vi. Demodulates the 16-QAM waveform. The excess points from the pulse shaping convolution are trimmed off, the waveform is downsampled and the recovered symbols are scaled.

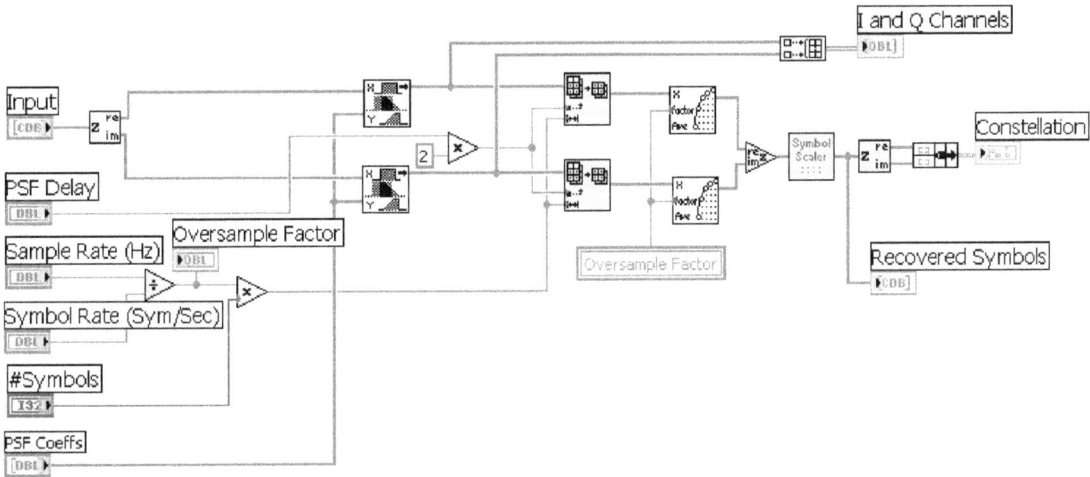

13. Downsample.vi. Lowers the sample rate of a discrete time waveform by keeping only one of every M samples. Filtering may be necessary before this operation to keep the expanded bandwidth of the downsampled sequence within the $f_s/2$ limits.

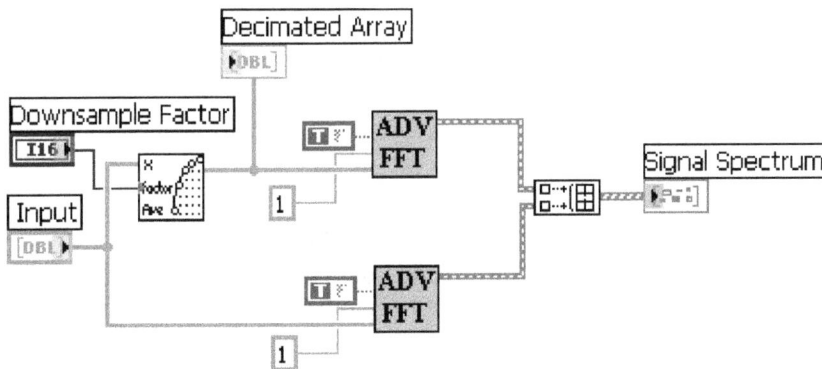

Downsample.vi keeps only 1 out of M samples of input.

14. EVM.vi. This VI calculates the error vector magnitude of a recovered symbol sequence by comparing it to a reference sequence. The magnitude of the errors is normalized to the magnitude of the constellation corner symbol.

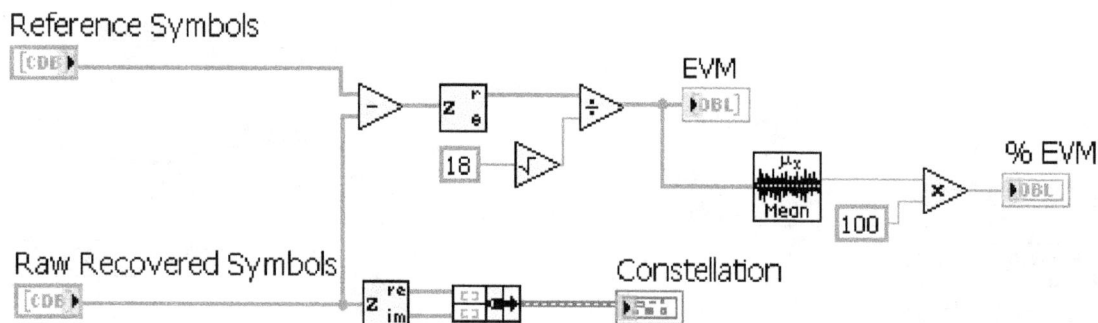

15. FadeSimulation.vi. Simulates a fading environment by generating a 16-QAM waveform, applying fading using Channel.vi, and demodulating the faded signal.

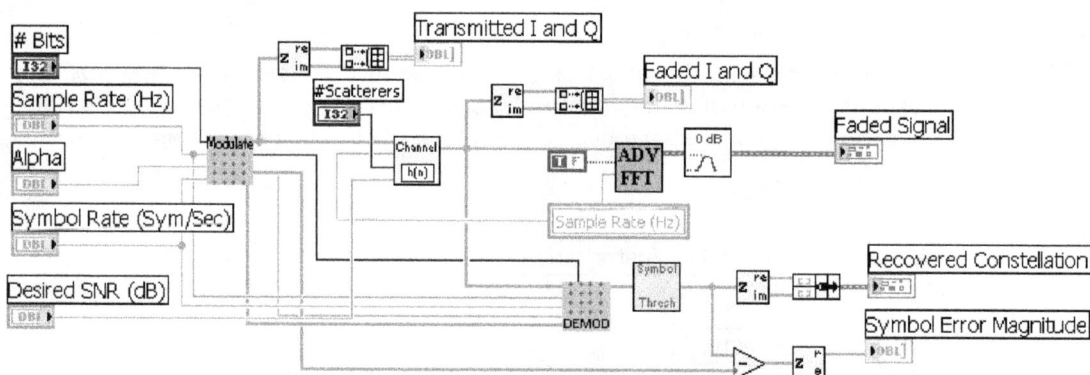

16. FastRealFFT.vi. Computes the FFT of a length $2N$ real sequence by reassembling it into an N-point complex sequence and using the complex FFT routine. The N-point complex spectrum is then operated on to produce the $2N$-point desired FFT.

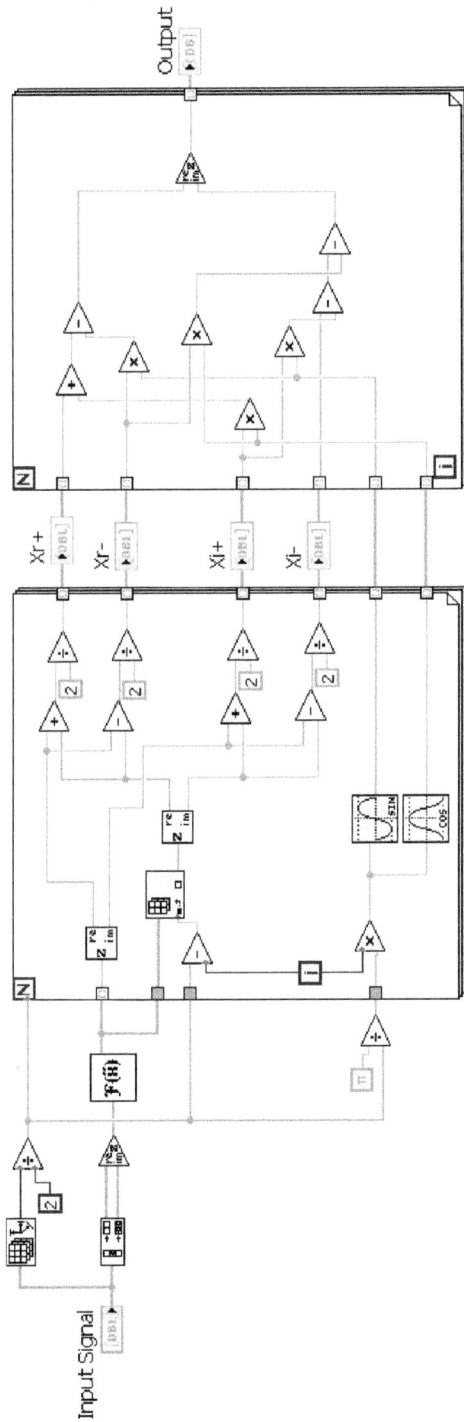

17. FreqAxis.vi.   Generates the frequency axis values for plotting a two-sided or single-sided spectrum.

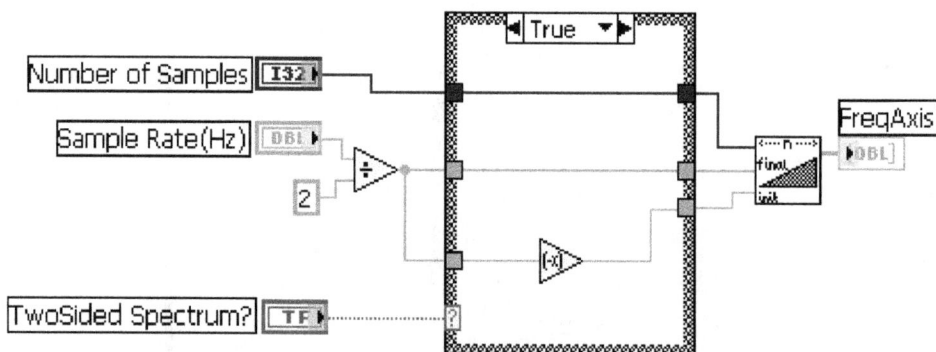

18. GenerateBits.vi.   Forms a sequence of bits using random number generator in a loop and rounding the result to the nearest integer.

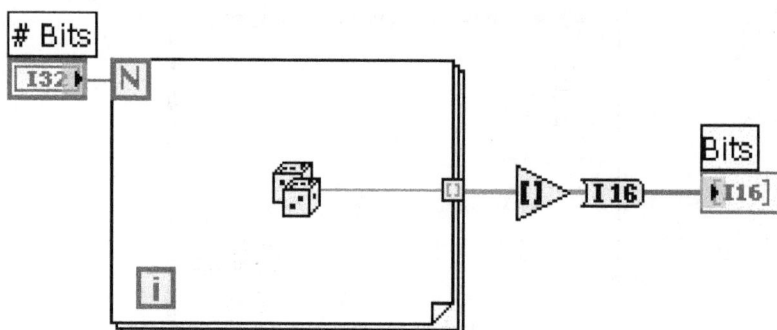

19. Halfband.vi.   Computes the filter coefficients for a lowpass filter (using the Parks-McClellan algorithm) with a cutoff of ¼ the sample rate. The halfband filter has the property of every other coefficient being zero.

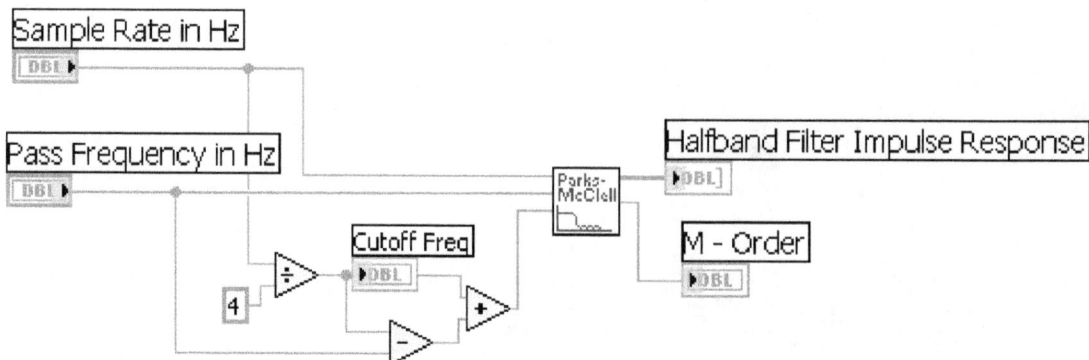

20. IIR.vi. Generates lowpass IIR filter coefficients using the Butterworth LabVIEW design VI. The filter order is computed from the design formulas.

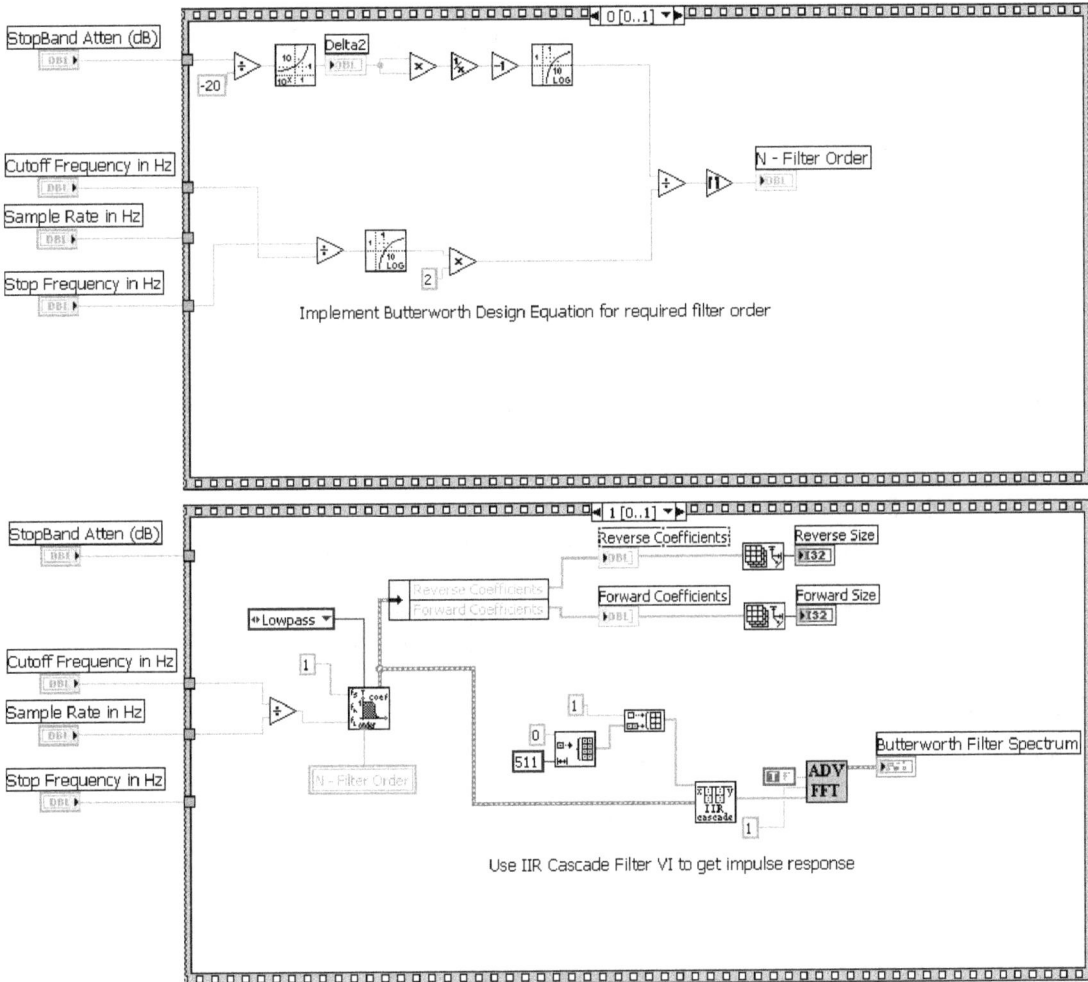

Implement Butterworth Design Equation for required filter order

Use IIR Cascade Filter VI to get impulse response

21. Init5660.vi. Initializes an instance of the PXI-5660 hardware. The taskID and scope handle are passed on to the read and close functions.

Initialize the Downconverter and Digitizer

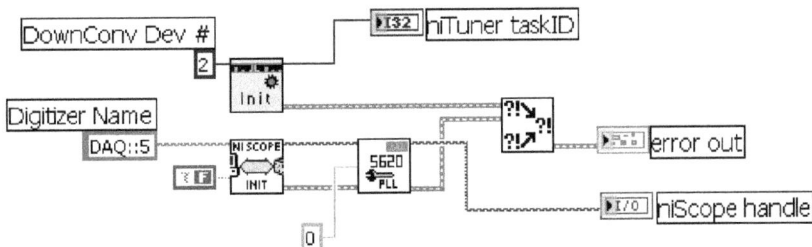

22. KaiserFIR.vi.   This VI generates digital filter coefficients based on the Kaiser window method.

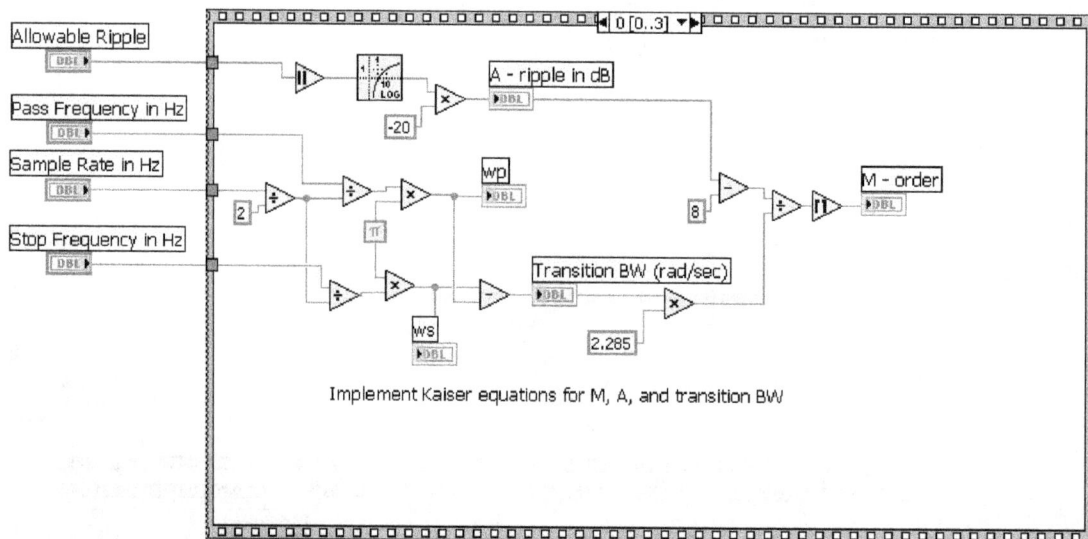

Allowable Ripple

Pass Frequency in Hz

Sample Rate in Hz

Stop Frequency in Hz

A - ripple in dB

wp

Transition BW (rad/sec)

ws

2.285

M - order

Implement Kaiser equations for M, A, and transition BW

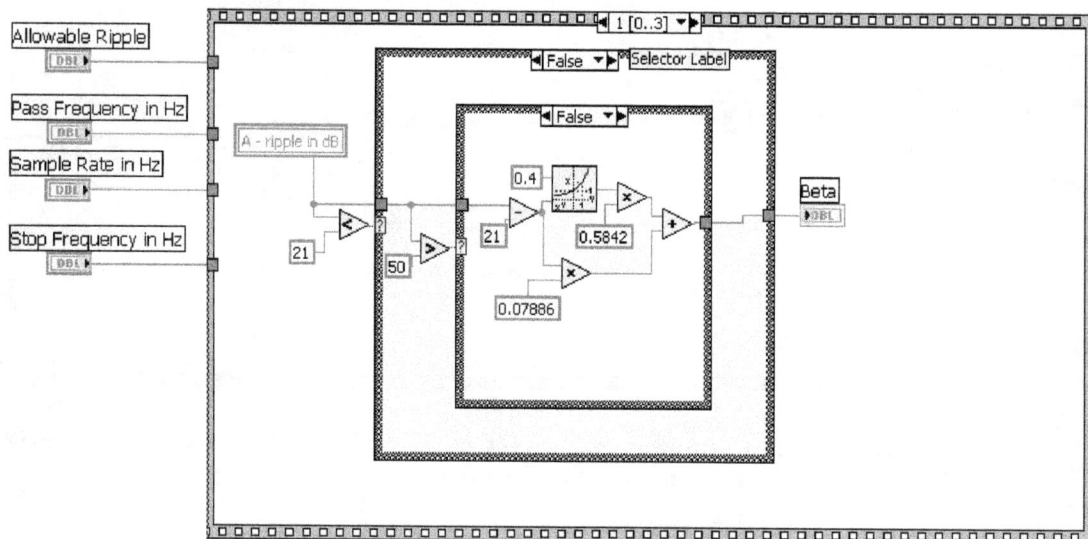

Allowable Ripple

Pass Frequency in Hz

Sample Rate in Hz

Stop Frequency in Hz

A - ripple in dB

Beta

False   Selector Label

False

21

50

21

0.4

0.5842

0.07886

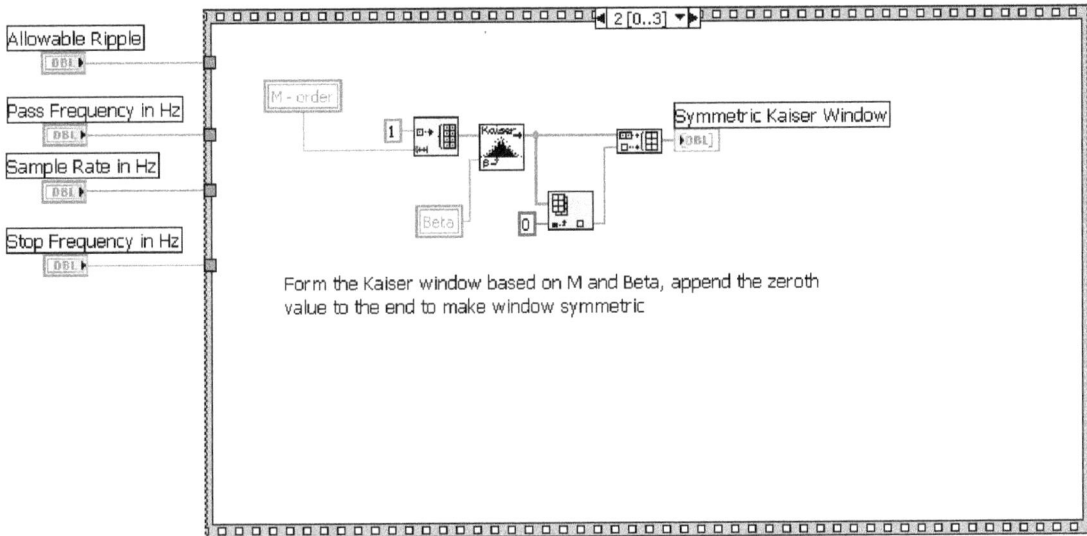

Form the Kaiser window based on M and Beta, append the zeroth
value to the end to make window symmetric

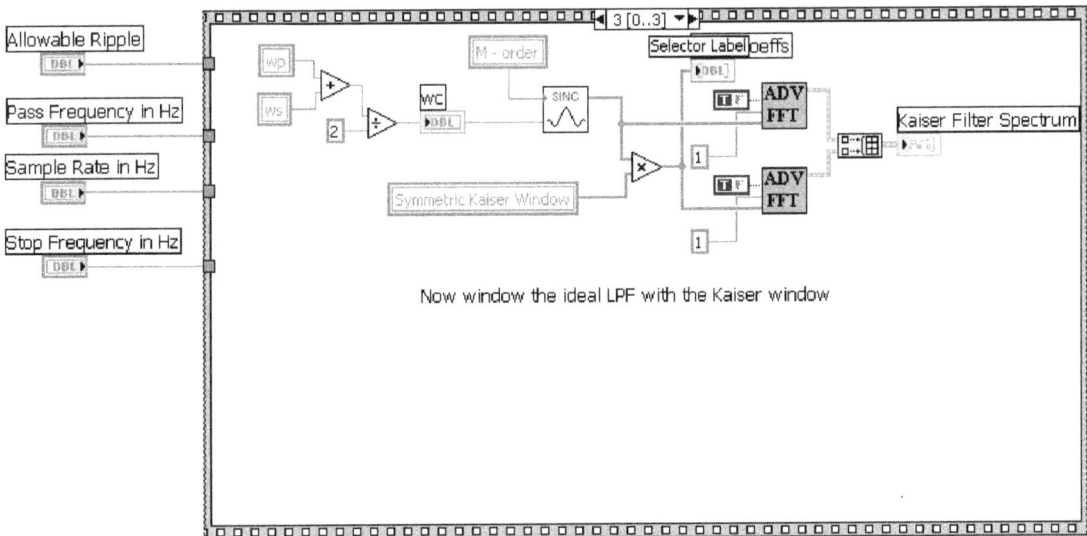

Now window the ideal LPF with the Kaiser window

23. LPCcoeffs.vi.   Computes the set of filter coefficients for linear predictive
model of speech segment. These coefficients can be used to synthesize a frame
of speech as an all-pole filter model.

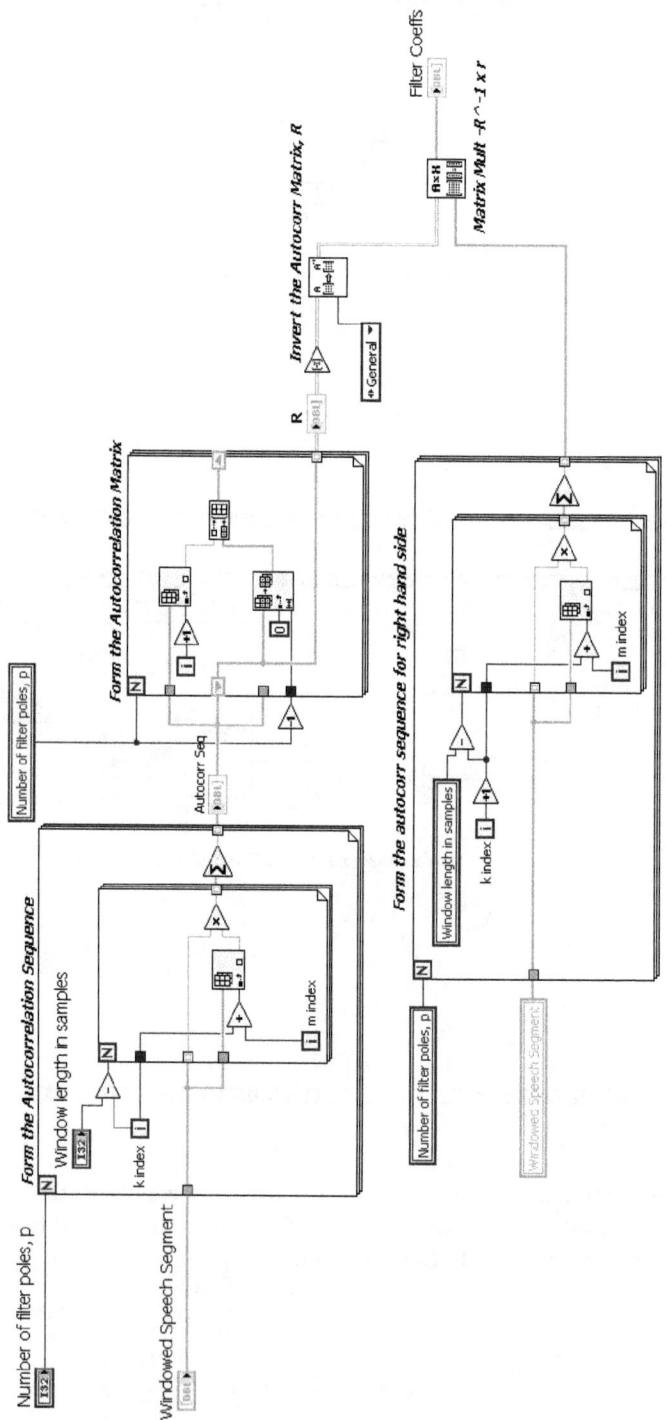

Number of filter poles, p

Windowed Speech Segment

Form the Autocorrelation Sequence

Window length in samples

k index

m index

Form the Autocorrelation Matrix

Number of filter poles, p

Autocorr Seq

R

Invert the Autocorr Matrix, R

General

Filter Coeffs

Matrix Mult -R^-1 x r

Form the autocorr sequence for right hand side

Number of filter poles, p

Window length in samples

k index

m index

Windowed Speech Segment

24. MatchedFilterDetection.vi. Demonstration of running a noisy input signal through a matched filter to determine which of the two signals was present.

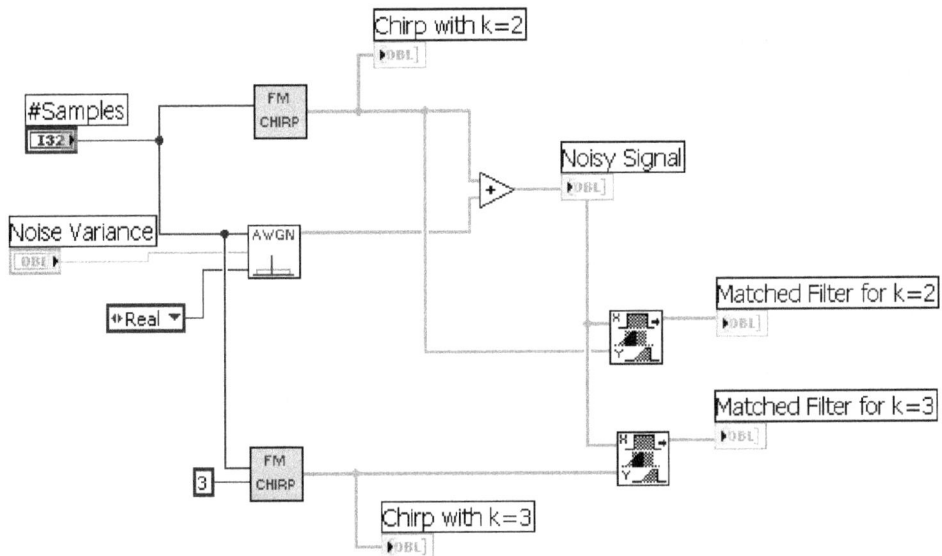

25. Modulator.vi. This VI modulates a sequence of bits by mapping them to a 16-QAM constellation, upsampling the symbol waveform, and finally shaping the pulses for zero ISI.

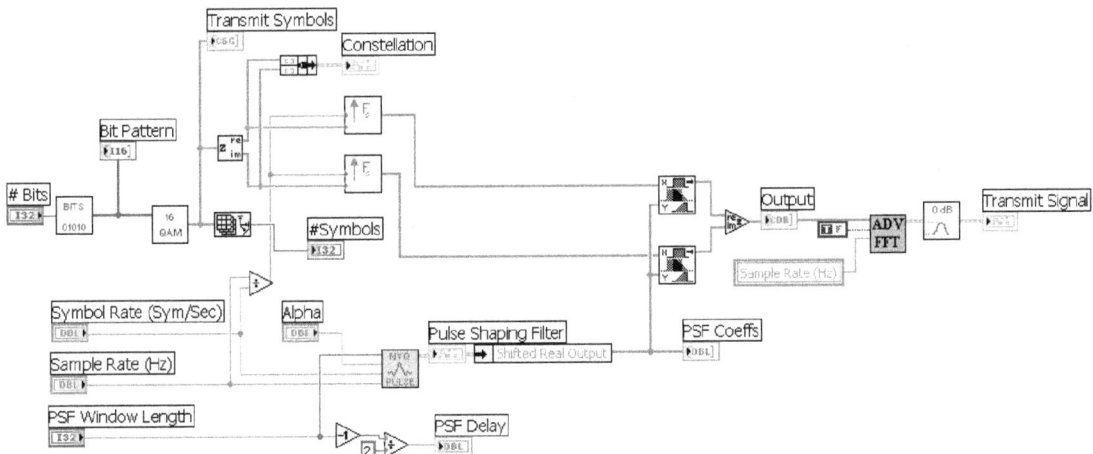

26. MPR.vi.   Generates lowpass filter coefficients using the optimum Parks-McClellan algorithm. This VI will choose the appropriate filter order or the user can specify a desired filter order.

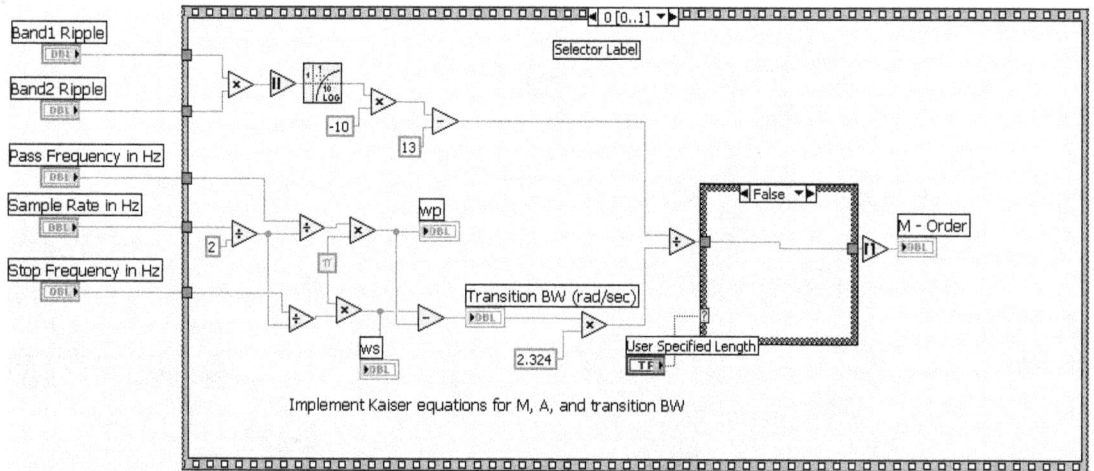

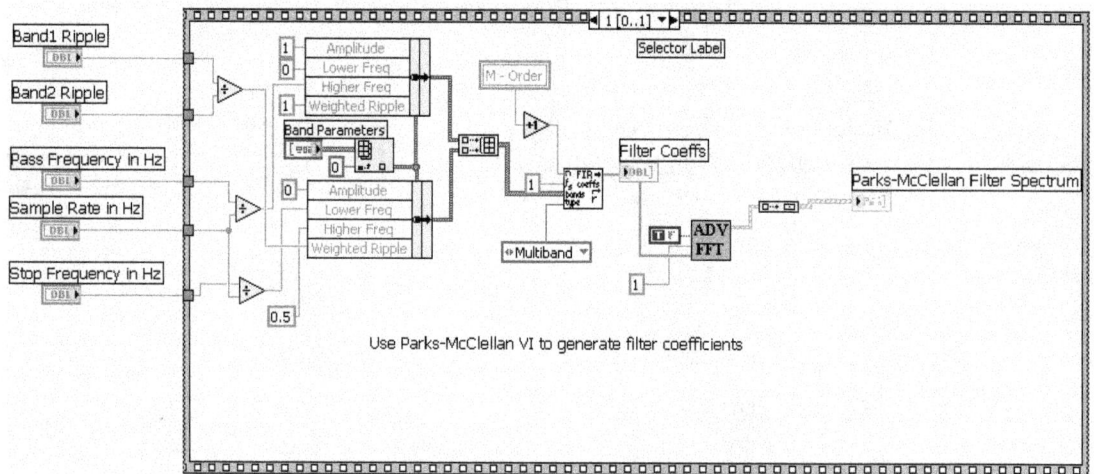

Implement Kaiser equations for M, A, and transition BW

Use Parks-McClellan VI to generate filter coefficients

27. NyquistPulse.vi.   Generates either a raised cosine or root-raised cosine pulse-shaping filter. The raised cosine filter is from the Nyquist class of shaping pulses, which exhibit zero intersymbol interference.

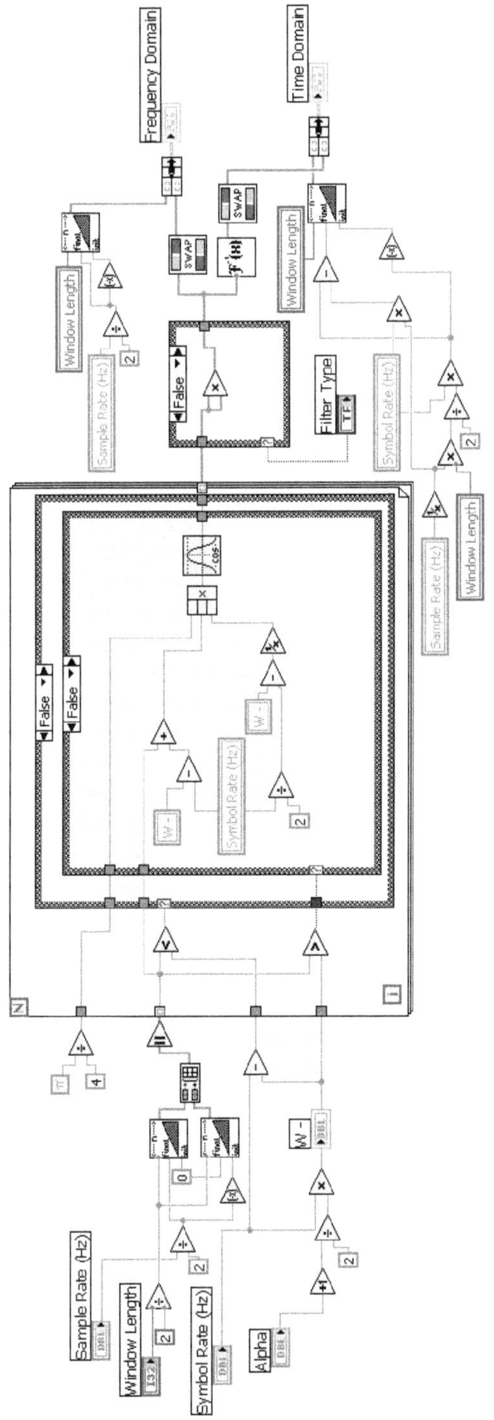
Frequency Domain
Time Domain
Window Length
Sample Rate (Hz)
Symbol Rate (Hz)
Filter Type
Alpha
False
False
False
SWAP
SWAP
W -

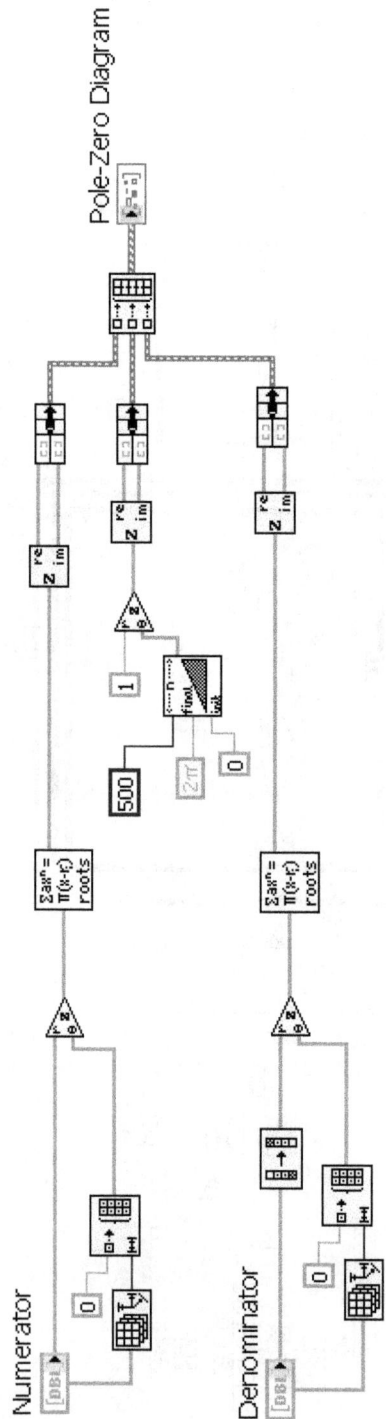
Pole-Zero Diagram

Numerator

Denominator

28. PoleZero.vi. Solves a linear constant coefficient difference equation and plots the poles and zeros in relation to the unit circle.

29. PolyphaseInterpolator.vi. Upsample and filter using efficient polyphase approach. The filter is split into separate banks and the input is upsampled and filtered in a combined operation.

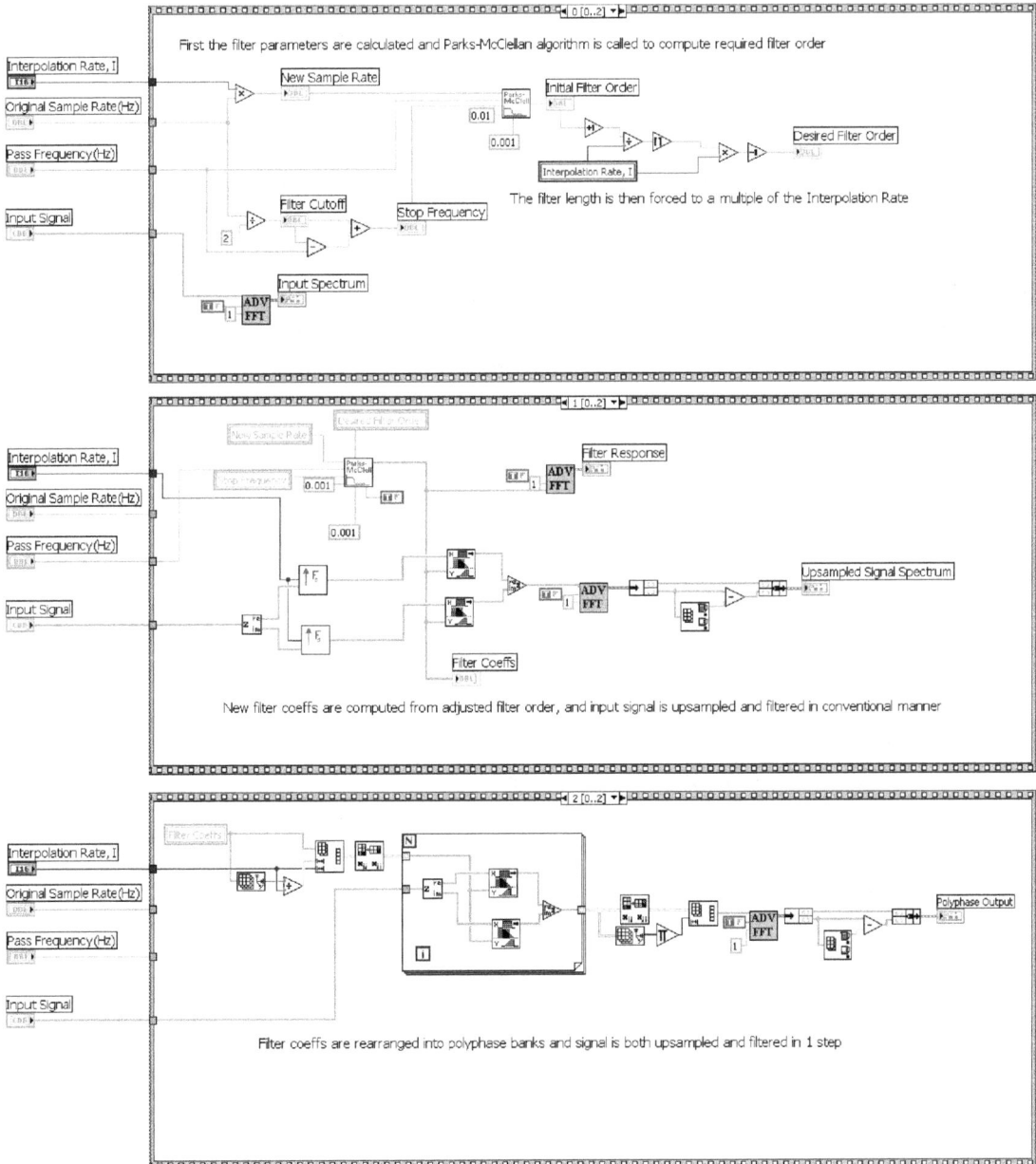

30. PulseFreq.vi.   Calculates the frequency values for the Nyquist pulse VI.

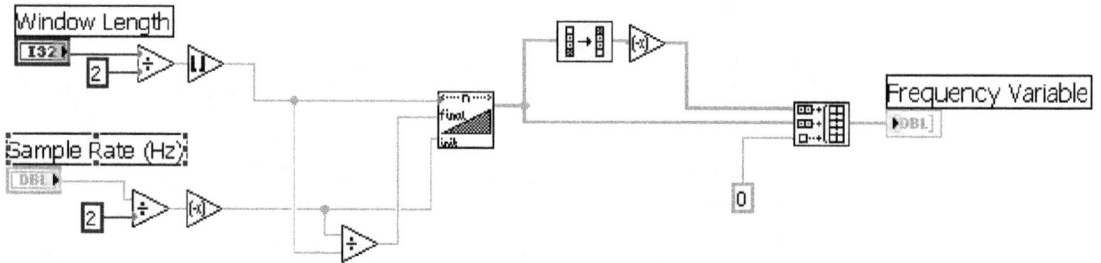

31. PXICapture.vi.   Utilizes the PXI-5660 downconverter and digitizer to continuously capture a record of specified time length.

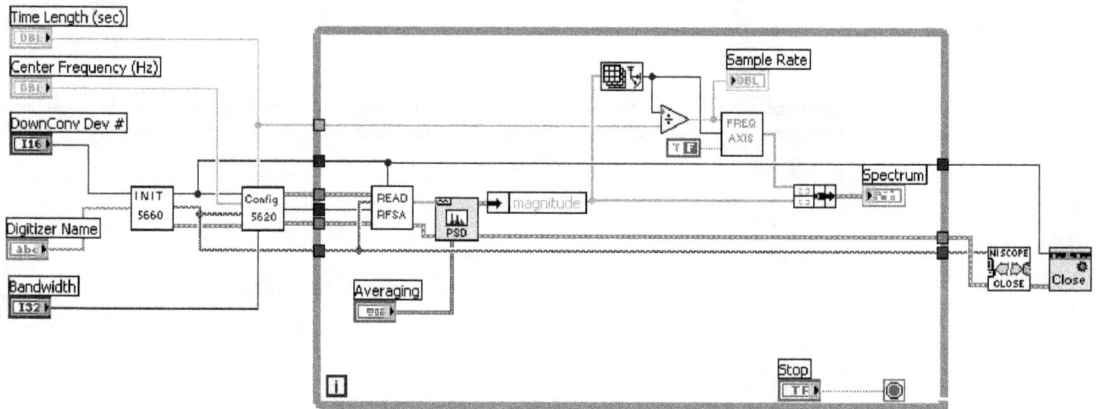

32. Rayleigh.vi.   Generates Rayleigh random variable using gaussian real and imaginary components and uniformly distributed phase.

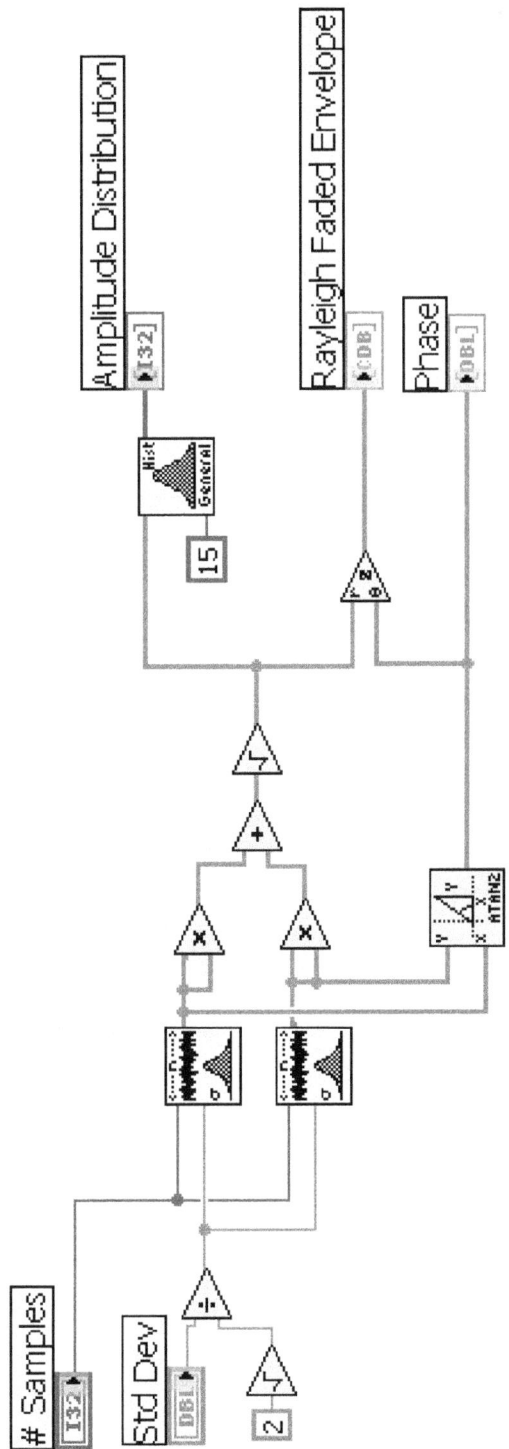

Phase is uniformly distributed from −pi to pi

33. ScaleSymbols.vi. Scales the recovered symbols so that the maximum magnitude is $\sqrt{18}$ for 16-QAM constellation.

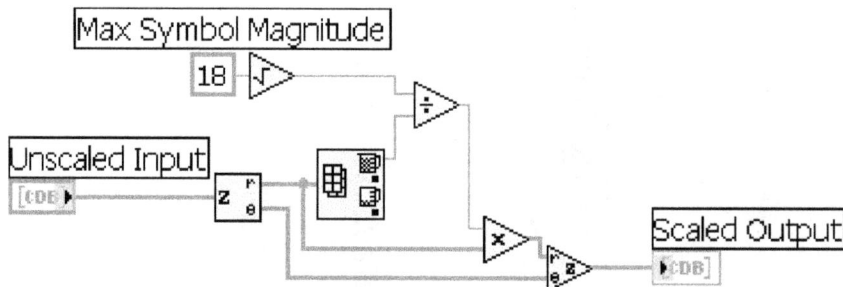

34. Set0dB.vi. Sets the peak value of a log magnitude plot to 0 dB.

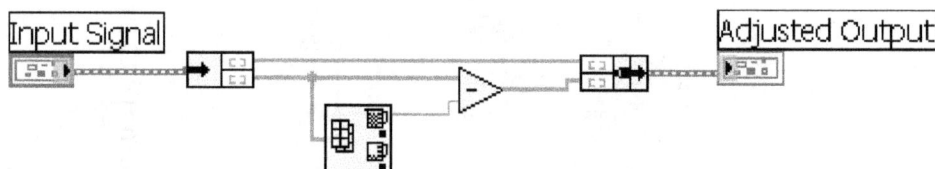

35. SimpleFFT.vi. Computes FFT with built-in LabVIEW function, converts magnitude to decibel, then shifts output to center at 0 Hz.

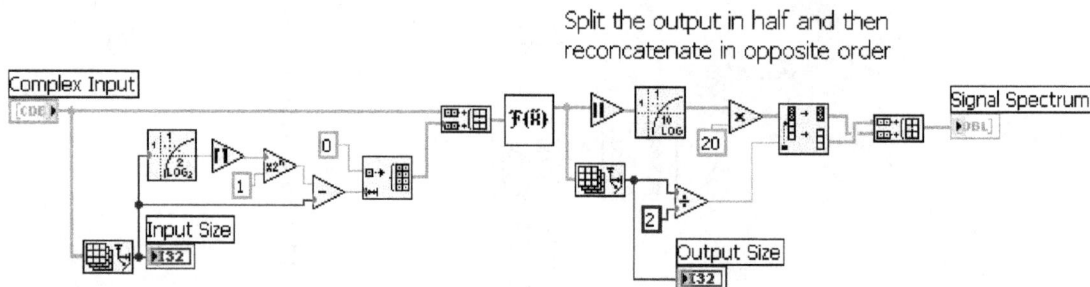

Split the output in half and then reconcatenate in opposite order

Here we bump the input size up to the next power of 2 size by appending zeros

36. SimpleNoisySystem.vi. Demo communication system, which calls the modulator, adds AWGN noise, and then demodulates. The recovered symbol constellation shows some scatter as a result of the noise.

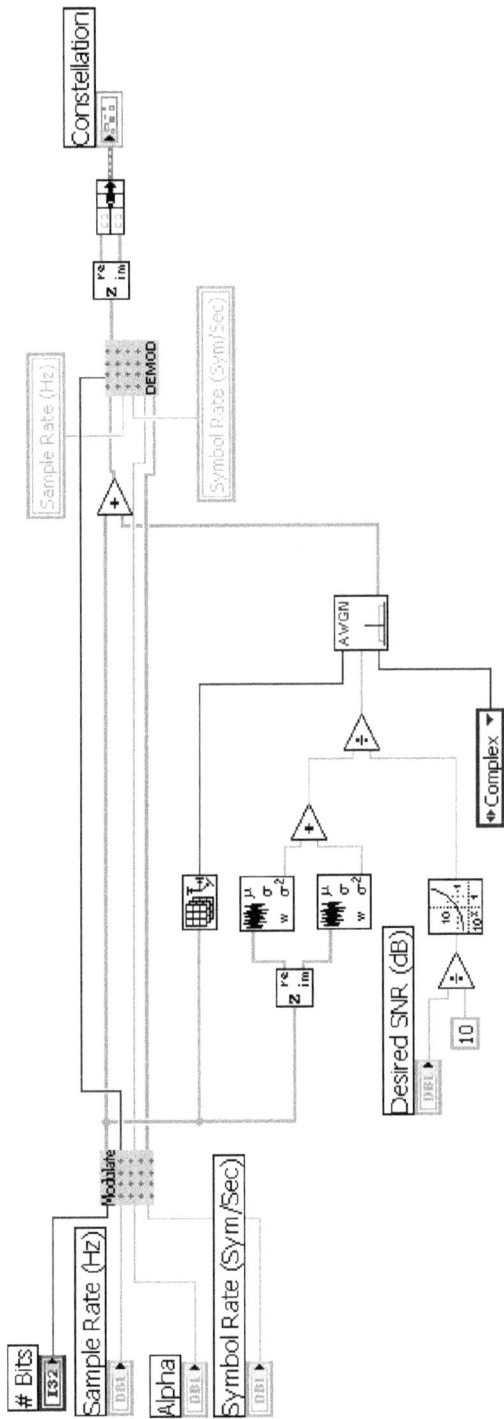

Length - number of samples for in sinc

Cutoff freq (rad/sec) - cutoff = (pass freq + stop freq) / 2

37. SincFcn.vi. Computes a discrete sinc waveform with a specified length and cutoff frequency. The cutoff is used because the sinc function forms a Fourier pair with the ideal lowpass filter.

38. SoundCardCapture.vi. Uses a PC soundcard to digitize analog input. The samples can be 8 or 16 bit, mono or stereo.

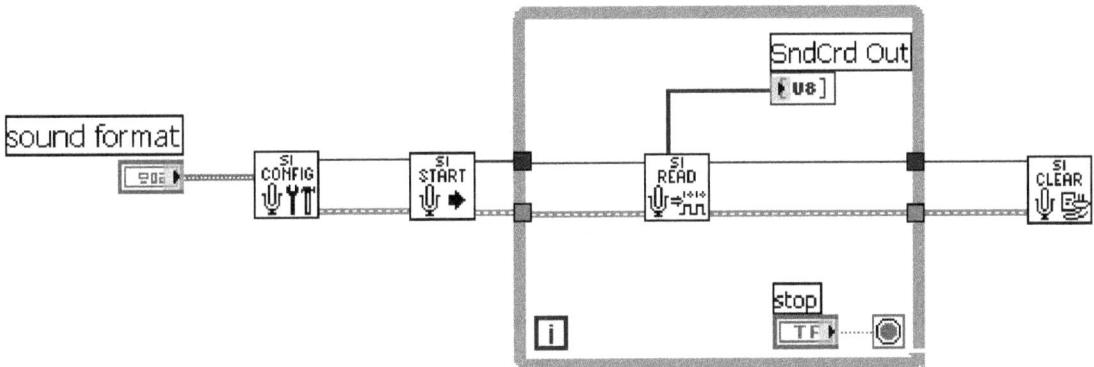

39. SymbolDecoder.vi. Simple VI to unmap the recovered symbols back into a sequence of bits.

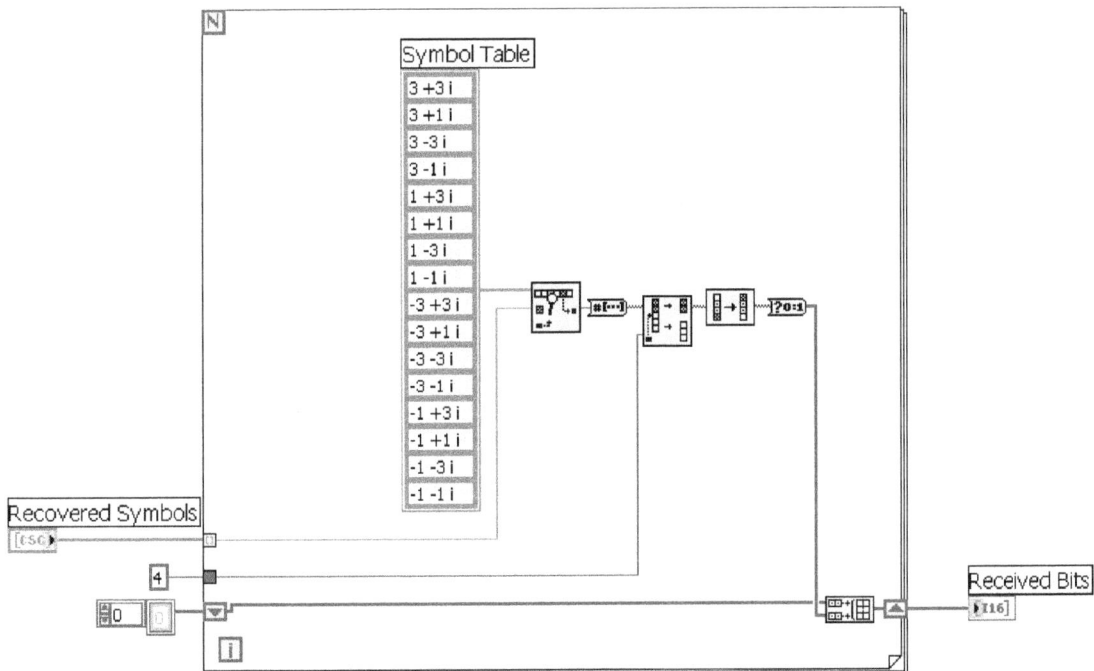

40. SymbolMapper.vi. Performs mapping of a bit sequence to points on the 16-QAM constellation.

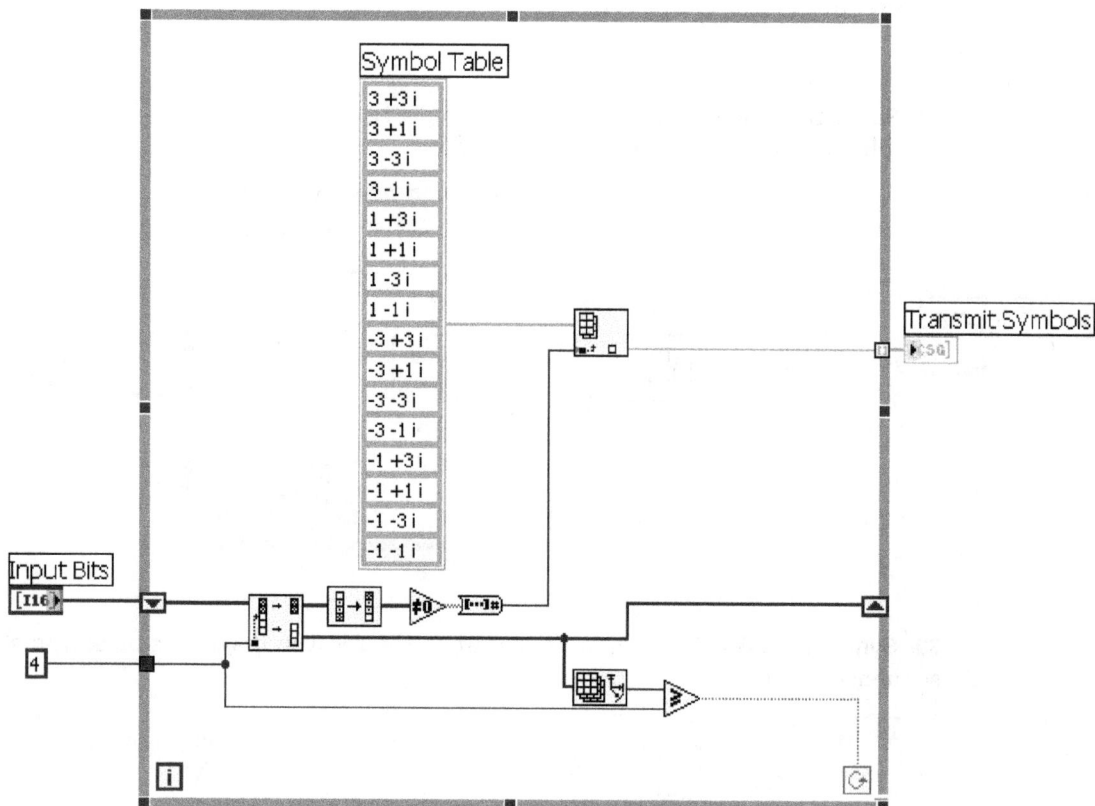

41. SymbolRateAxis.vi. Forms the frequency axis for the pulse-shape filter display. The frequency values are normalized to the symbol rate.

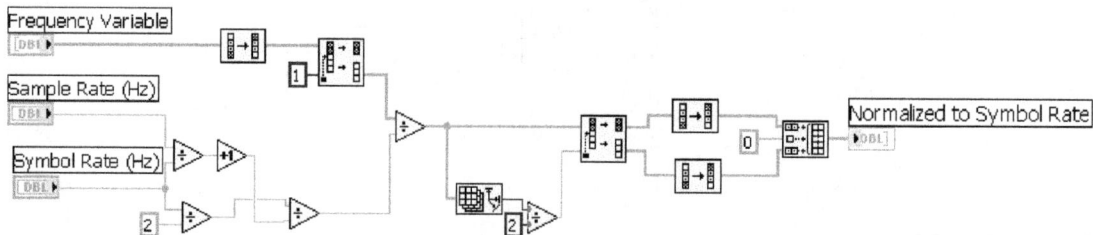

42. TrellisCodedModulator.vi. Performs trellis-coded modulation on specified number of bits. TrellisStateMachine.vi is called to compute current output symbol index as well as the next state.

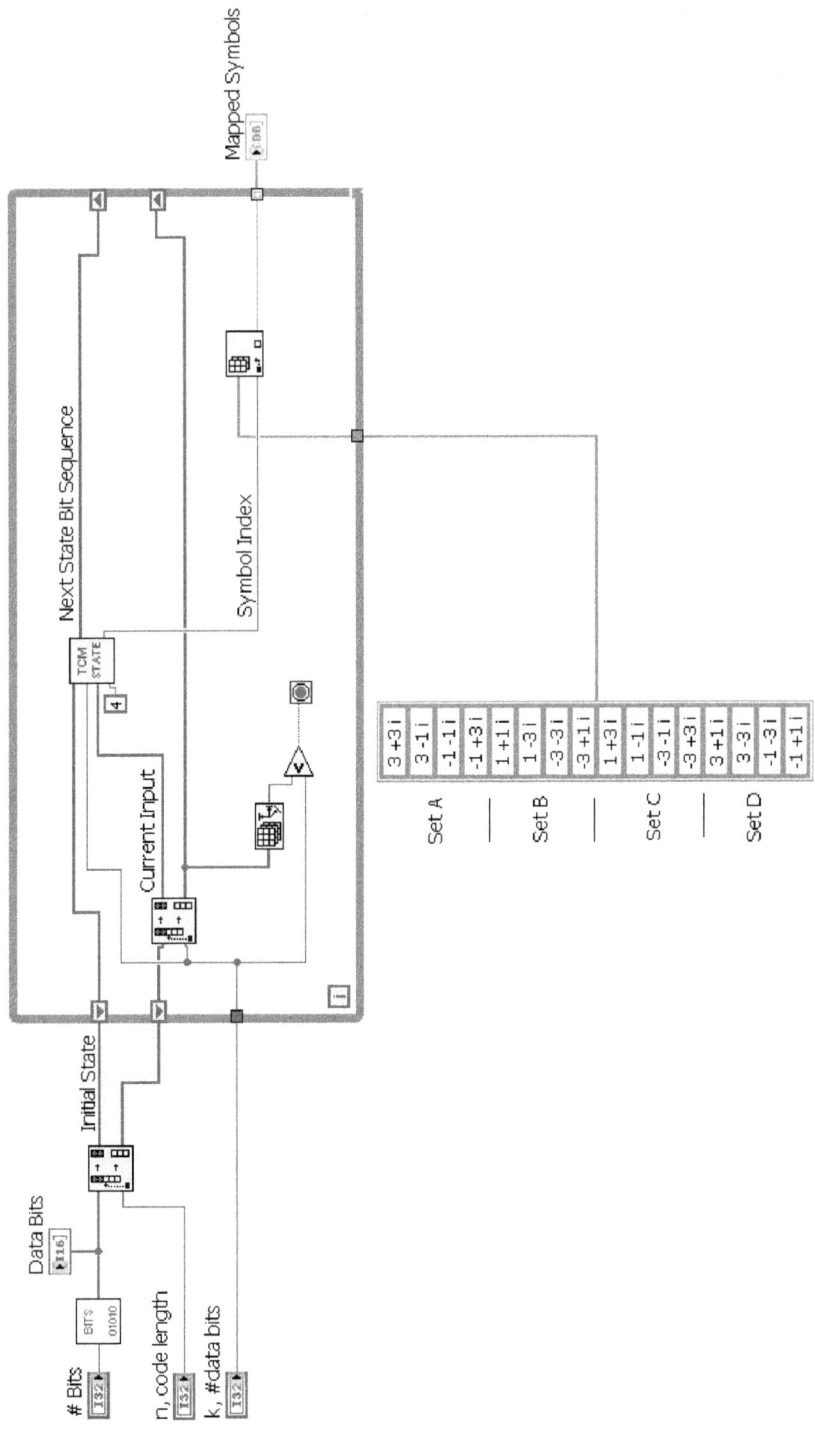

Dragen

Mapped Symbols

Next State Bit Sequence

Symbol Index

TCM STATE

4

Current Input

Initial State

Data Bits

Bits

BITS 01010

n, code length

k, #data bits

Set A	3 +3i
	3 -1i
	-1 -1i
	-1 +3i
Set B	1 +1i
	1 -3i
	-3 -3i
	-3 +1i
Set C	1 +3i
	1 -1i
	-3 -1i
	-3 +3i
Set D	3 +1i
	3 -3i
	-1 -3i
	-1 +1i

43. TrellisStateMachine.vi.   Chooses the next state and current output symbols based on the current state and current input. This VI is then used in a trellis-coded modulator.

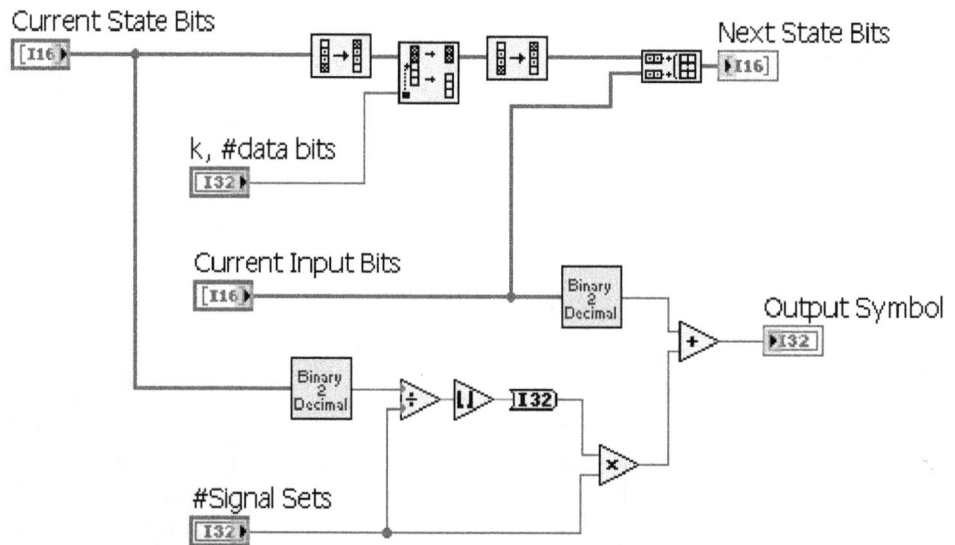

44. Upsample.vi.   Upsamples the real input sequence by stuffing upsample-factor-minus-one zeros in between each sample. There is no filtering done on the output.

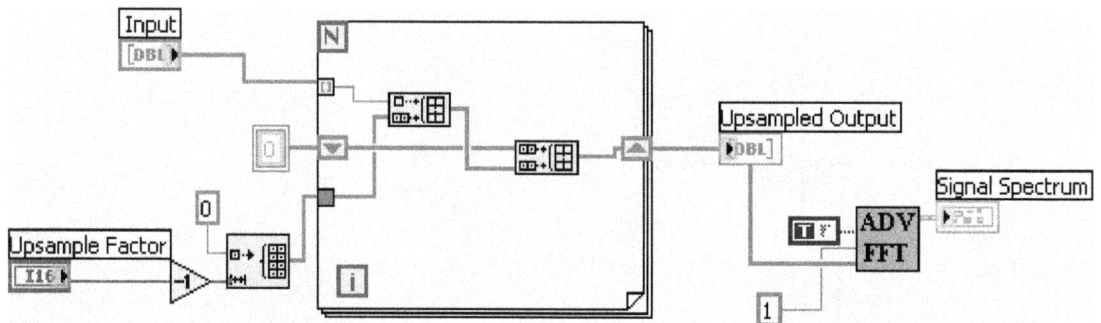

Upsample.vi stuffs zeros between each sample of the input signal.

45. Viterbi_distance.vi. Computes the Euclidean distance between a raw recovered symbol and the closest matching symbol in each signal set. Not a complete Viterbi decoder, this VI simply computes the distance metric for each symbol.

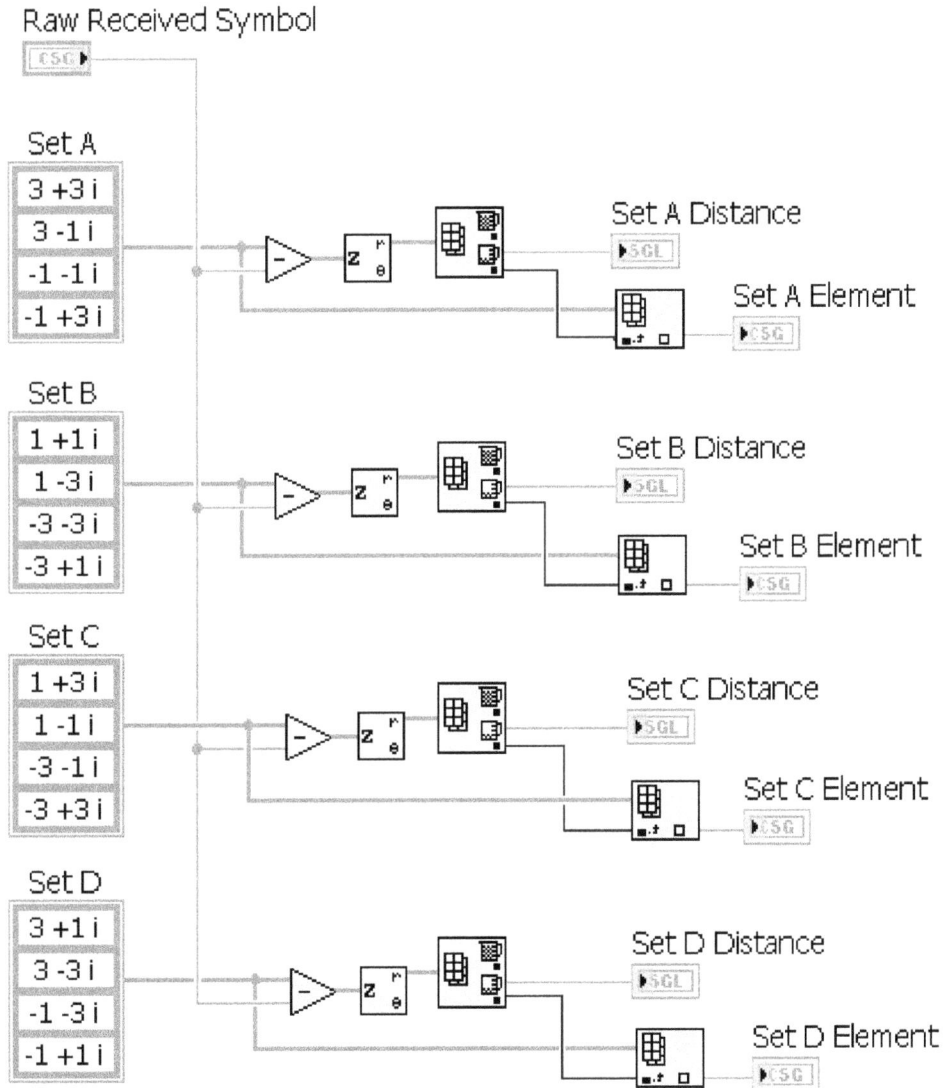

Raw Received Symbol

Set A

| 3 +3 i |
| 3 -1 i |
| -1 -1 i |
| -1 +3 i |

Set A Distance

Set A Element

Set B

| 1 +1 i |
| 1 -3 i |
| -3 -3 i |
| -3 +1 i |

Set B Distance

Set B Element

Set C

| 1 +3 i |
| 1 -1 i |
| -3 -1 i |
| -3 +3 i |

Set C Distance

Set C Element

Set D

| 3 +1 i |
| 3 -3 i |
| -1 -3 i |
| -1 +1 i |

Set D Distance

Set D Element

46. Window.vi.   Demonstrates the FFT of a signal with and without windowing. The window function can be chosen from a dropdown box.

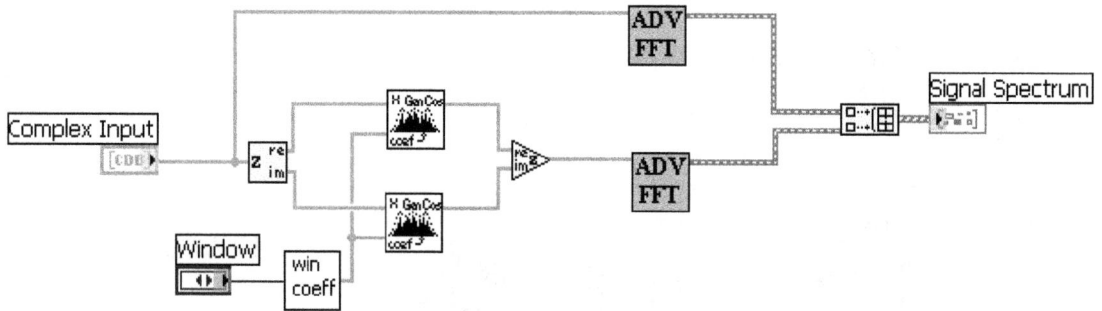

Hardware Resources

There is not a lot of sampling hardware available that will operate at the frequencies that most digital communication devices use. The few companies listed here all have some very good products suitable for the type of work mentioned in this book. Hopefully there will be more additions to this list in the near future; for now start with these companies.

National Instruments (*www.ni.com*). NI offers a large selection of sampling hardware, although most of the products are not directly suitable for RF frequencies. The largest bandwidth digitizer appears to be the PCI-5112. This 8-bit digitizer has 100 MHz of analog bandwidth and can sample up to 100 Msps—not quite enough for most RF signals. More interesting for digital communication work is the PXI-5660 (includes the PXI-5620 digitizer and PXI-5600 RF downconverter). The 5660 is billed as an RF signal analyzer and this device was explained in some detail in Chap. 2. The 5600 RF downconverter has a bandwidth of up to 2 GHz. It also ships with some nice LabVIEW tools for spectral analysis in the form of the spectral measurements toolset. NI has recently added the PXI-5670 to their RF lineup. The 5670 combines an arbitrary waveform generator with an RF signal source to now offer the capability of generating digitally modulated RF signals. Both the 5660 and 5670 are only available in the PXI form-factor and therefore require a PXI chassis, power supply, and one of the controller options mentioned in Chap. 2.

Acqiris (*www.acqiris.com*). Acqiris offers an entire family of 8-bit PCI digitizers with input bandwidths from 150 MHz up to 1 GHz. These cards also ship with complete LabVIEW support in the form of a library accessible from the functions palette as shown in Chap. 2. The product is simple to use, comes with some example virtual instruments (VIs), and the LabVIEW interface is explained in the manual. This product was used in conjunction with the subsampling technique discussed in Chaps. 1 and 2 and performed as expected. The install and interface were smooth and the card was up and running in under 20 min.

Gage Applied Technologies (*www.gage-applied.com*).   Like Acqiris, Gage also has a line of 8-bit PCI digitizers and they also offer up to 1 GHz of bandwidth. Gage also lists LabVIEW support on their Web page, although this has not been confirmed first hand. It is also worth mentioning that Gage sells oscilloscopelike software for the PC, which will acquire and display data from their digitizers.

Delphi Engineering (*www.delphieng.com*).   Delphi offers some extremely high bandwidth products, specifically the ADC3200 (10-bit) and ADC3100 (8-bit). These are both PMC modules with up to 3 GHz (ADC3200) analog bandwidth and sample rates up to 2 GHz. The downside is that these are PMC modules, but there are PMC to PCI converters available that allow you to plug this into a standard PCI slot.

Index

ABOUT THE AUTHOR

Cory L. Clark is a senior software engineer with Motorola and has developed many LabVIEW-based DSP tools. He holds a master's degree in electrical engineering from Georgia Tech.

www.ingramcontent.com/pod-product-compliance
Lightning Source LLC
Chambersburg PA
CBHW061923190326
41458CB00009B/2637